Stephen Forbes

AND
THE
RISE
OF
AMERICAN
ECOLOGY

Stephen Forbes

AND

THE

RISE

OF

AMERICAN

ECOLOGY

ROBERT A.

CROKER

SMITHSONIAN INSTITUTION PRESS | WASHINGTON AND LONDON

Publication of this book was supported by a grant from the
Illinois Natural History Survey.

Editor: Jack Kirshbaum
Production editor: Ruth Spiegel
Designer: Linda McKnight

Library of Congress Cataloging-in-Publication Data
Croker, Robert A.
   Stephen Forbes and the rise of American ecology /
Robert A. Croker
      p.   cm.
   Includes bibliographical references (p.)
   ISBN 1-56098-972-6 (alk. paper)
   1. Forbes, Stephen Alfred, 1844–1930. 2. Biologist—United
States—Biography. 3. Ecologists—United States—
Biography. 4. Ecology—United States—History. I. Title.
QH31.F62 C76 2001
508'.92—dc21
[B]   2001018377

♾ The paper used in this publication meets the minimum re-
quirements of the American National Standard for
Information Sciences—Permanence of Paper for Printed
Library Materials ANSI Z39.48-1984.

Contents

Preface

Like many a student, I stumbled upon Stephen Forbes's seminal 1887 essay, "The Lake as a Microcosm" while reading for a graduate course in ecology. Immediately engaging, it resonated far ahead of its time and was clear by Forbes's foundational synthesis that it described what we later termed the "ecosystem." Years later I learned more of Forbes's connections with the Illinois Natural History Survey and the Illinois River—"the best-studied stream in the world"—and became more broadly aware of his celebrated work in natural history and early ecology. My interest in Forbes has now spanned more than forty years, leading me to explore his ancestry, youth, and private and professional lives.

From afar his life was similar to thousands of others in the nineteenth century who were raised on the midwestern American frontier. And like so many of his generation, he was shaped by hard times and his powerful experience as a Civil War soldier. In postwar years Forbes's life took quite a turn, and on the way Forbes reported being swept along a course he did not fully understand, but whose destination he grew to trust more and more. More than most, he had mastered the art of living with uncertainty.

Liberal learning served as his early compass—classical literature, languages, philosophy, and history, as well as the Methodist faith. Yet he never allowed any special school or creed to hem in his freedom of thought or action.

Forbes was in the forefront of biological work while ecology emerged and formed as a recognizable scientific discipline. I wrote this book to develop the character of the man as he practiced his science, science which was always in the public service. It was this commitment that guided Forbes in his discovery and development of the concept of ecological systems and in his pioneering work toward a truly environmental science.

My warmest thanks for insight into the Forbes story goes to Stephen Forbes's grandson, Richard M. Forbes, professor emeritus of nutritional biochemistry, University of Illinois, and his wife, Mary. Their assistance, encouragement, and balanced view of Stephen Forbes and his legacy spearheaded this book. Meeting and talking with their adult children, Anne, Sally, and Stephen, about their great-grandfather was memorable. Two other family members, the late Mary (Scott) Cox and Thomas W. Scott, provided numerous colorful letters, recollections, and photographs of their grandparents.

Several people graciously gave noteworthy assistance: John Hoffmann, archivist, Illinois Historical Survey, University of Illinois Library, guided me through the Stephen Alfred Forbes Collection; Robert T. Chapel, technical assistant, University of Illinois Archives, helped with Forbes's voluminous correspondence and other papers; Beth Wohlgemuth, JoAnn Jacoby, Erin Knight, and Monica Lusk, librarians at the Illinois Natural History Survey, assisted with Forbes's scientific papers; Thomas Rice, technical editor, Illinois Natural History Survey, searched for salient photos and discovered unpublished notes on Robert E. Richardson; Steven Havera, director of the Forbes Biological Station, and his indefatigable assistant, Katie Roat, Illinois Natural History Survey, Havana, Illinois, hosted my wife's and my visit to the station and provided access to Forbes materials there; the late Lt. Col. Albert J. Sambold (U.S. Air Force) made several splendid prints from old pictures; and Ron Bergeron, photographer, Information Services, University of New Hampshire, supplied excellent prints of Forbes family members.

Several colleagues and scholars should be singled out for helpful correspondence and conversations: Eldon Johnson, Rob Wiedenmann, Rick Weinzierl, Edward Smith, Stephen Bocking, Robert P. McIntosh, Steven Havera, and the late Robert L. Metcalf. The latter two men also read parts of the manuscript and shared their comments. Richard Forbes and two anonymous reviewers read the entire manuscript. The constructive suggestions of all these people were sincerely appreciated. Special regard goes to Vincent Burke, science editor at Smithsonian Institution Press, for his guidance and encouragement throughout the book-publishing journey, and to Jack Kirshbaum for his astute and careful editing.

I am indebted to the Illinois Natural History Survey, Dr. David Thomas, chief, for a generous grant in support of publication of this book.

Many other individuals provided assistance at the following institutions: the Chicago Historical Society; the Rush-Presbyterian–St. Luke's Medical Center

Archives in Chicago, the U.S. National Archives & Records Administration; the National Academy of Sciences; the U.S. Army Military History Institute in Carlisle Barracks, Pennsylvania; special collections at Bradley University Library in Peoria, Illinois, and Bowdoin College, Brunswick, Maine; the Stephenson County Genealogical Society, and Land Record Office, and the Freeport Public Library, all in Freeport, Illinois; and the University of New Hampshire Libraries in Durham. Teresa McAlhany, Erin Dauphinais, and Ellen Bullard expertly typed several versions in manuscript.

I began this book in 1990 while a faculty member in the Department of Natural Resources, University of New Hampshire, including a fall semester sabbatical leave in 1993. I wrote the large majority of the book, however, after my 1994 retirement from the university. My wife of twenty-nine years, Mary Ann, deserves my thanks and affection for her help with various manuscript details, her patience with my "writing moods," her loyalty, and her having me not only for love, but for lunch this past half-dozen years.

Stephen Forbes

AND
THE
RISE
OF
AMERICAN
ECOLOGY

Prologue

Late in the afternoon of Friday, February 25, 1887, an erect, handsome, athletic-looking man of 42 stepped lively from a train in Peoria, Illinois. His bearing suggested an attitude of command. In fact, twenty-two years earlier he had been one of the youngest captains in the western Union Army, commanding Company B, Seventh Illinois Volunteer Cavalry Regiment. Dr. Stephen Alfred Forbes picked up his bag and strode briskly up the rising ground to the west, where the city lay on a wide bluff overlooking a broad expanse of the Illinois River.

Forbes was at this time a professor of both zoology and entomology at the University of Illinois, the director of the Illinois State Laboratory of Natural History, and the state entomologist. Moreover, in a year he would take on another responsibility as dean of the College of Natural Science. Since 1876 he had worked on and published an extraordinary variety of natural history and zoological subjects, most with an ecological slant, though the word "ecology" had not yet caught on in the United States.

That February night in Peoria, Forbes would eloquently describe his concept of a microcosm, or ecological system, a statement that would later be considered the classic theoretical scheme from America for the study of lakes both here and abroad and a prelude to his formal recognition of the new science of ecology in 1894, one of the first U.S. zoologists to do so.

Furthermore, Forbes's remarkably prophetic and sophisticated work anticipated a dozen or more important concepts not addressed by ecologists until later in the twentieth century. Through his combining of ecological theory with practical applications of his science before and after the turn of the twentieth century, Forbes also laid an American foundation for environmental science in the public interest. Whether in the army at age 17 to 21, or as a fledgling physician, or later as a scientist, public service was the motto embroidered on the personal flag he flew for almost sixty years.

The central, defining experience of his life, however, was the crucible of his Civil War cavalry service, where he fought in two dozen skirmishes and battles, including the famous 1863 Grierson Raid deep into the heart of the Confederacy. Brave and fiercely patriotic, he proudly served under his older brother, who was his company commander and later his regimental colonel. In the cavalry he learned about himself and about leadership and dedication, organization and resourcefulness—skills that would contribute to his becoming the most widely accomplished ecologist of the nineteenth century. War, he said, was his early education. Yet he faithfully kept the flame burning for his wide-ranging intellectual interests with his saddlebag soldier's library, even while in four-month captivity as a Confederate prisoner of war. Something of the war's effect on him, his public spirit, and his lifelong love of nature come across in a letter to his sister Nettie, in 1883, almost twenty years after his mustering out of the army:

> I suppose that you will expect to hear something of my New Orleans trip? Or did you know that I went down? Two things impressed me most vividly. One was the long look over the [familiar] country in the dawn of the morning, from the platform at Grand Junction, Tennessee, my head ringing with bugle calls and the boom of cannon all the while. How strange it seemed to watch myself trotting along with jingling saber in that blue column of cavalry just out of sight behind the woods! What do the young men of this generation find to take the place of that overmastering, inspiring, devoted enthusiasm of the war time?—I haven't been good for anything since, that hadn't some seeming of public service about it.
>
> The other thing which I shall remember was, the extremely foreign—, the delightfully novel effect of a Louisiana swamp viewed from hip boots, or from the surface of a winding bayou. I don't think that anyone but a field naturalist can appreciate fully the stimulus of such novel surroundings. I tried to get home a tangible hint of the situation;—a fan palm, but that was too large to pack; a lovely Amaryllis which lifted its large white blossom deep in the lonely bog, but the delicate thing withered in an hour. . . . Of course the children were favored with the inevitable little alligator, which I brought home in a cigar box. They kept him until he was no longer horrible, and then gave him away.

To get steady bearings on the man, his ancestors, and his boyhood world of a youthful, expansive America, we must break into a story that unfolded even earlier. Three years after the American Revolution, army veteran, Stephen

Forbush (soon changed to Forbes), his wife, Mary, and their eight children, including ten-year-old John, Stephen Forbes's grandfather, left Hardwick, Massachusetts, headed northwest, and crossed the Deerfield River. They sought Stephen's veteran's land grant in the mountains of "New Connecticut," soon to become the village of Wilmington in the young state of Vermont. Each succeeding generation of the Forbes family would now trek optimistically further west.

THE

EARLY

YEARS

CHAPTER 1

Beginnings, 1836–60

Now it was their turn as he had promised. There on the autumn wind-swept prairie beyond the western sky curled that welcome stretch of the old army trail to journey's end.

So did Isaac and Agnes Forbes, onetime residents of Preble, New York, leave the riverside settlement of Aux Plaines near Chicago in October 1836, with their four young children, Flavilla, Francis, Mary, and Henry. Traveling by covered wagon and ox team, they headed northwest for a land claim on the rolling Illinois tall grass prairie 26 km (16 mi.) south of the Wisconsin border near the Pecatonica River.

Forbes and many others had heard favorable reports about northern Illinois and flocked westward, intent on getting fertile land. Indeed, 1836 was the peak of land claims in the history of the United States.

Isaac Forbes's pioneering had begun seven years earlier in 1829 to "see how things went in other parts." He and his older brother, Stephen Van Rensselear Forbes, left Preble and traveled west with a surveying party to Fort Dearborn. After exploring as far south as Louisiana, Isaac returned home to western New York. Stephen V. R. Forbes and his wife, Elvira, however, left Ohio and settled at Fort Dearborn in 1830 just as this frontier post was being transformed and renamed Chicago. Stephen was the first schoolteacher in Chicago. In a log house at the foot of what is now Michigan Avenue on the outlet of the Chicago River, he and Elvira taught both white and American Indian children. Two years later he was elected the first sheriff of Cook County, Illinois.

They soon built a large log house, and laid out a farm on 160 acres of land on the present site of Riverside, Illinois, in and around the great bend of the

Des Plaines River. Their land included what came to be known as Bourbon Springs, the adjacent island, the winding river east of the dam, the old mill site and Indian ford, and the fine wooded area of what was later called River-side Lawn. Their pioneer farm home, with its heavily timbered kitchen and huge fireplace, had warmly welcomed several dozen migrating Forbeses from Preble, New York, in the spring of 1836, including Isaac's own family and Isaac's parents, John and Anne Sawyer Forbes. It was from this log house, from the marshes, woods, and orchards surrounding it, that Agnes and Isaac's family left that fall for a new home and a fresh start on the Illinois frontier.[1]

They followed the U.S. Army trail blazed to Galena in 1832 during the Black Hawk War, over mile after mile of deep black arable soil, and through occasional isolated groves or streamside stands of hickory, black walnut, white oak, butternut, elm, and poplar. Within a fortnight they entered the endless sea of prairie flora such as bluestem, Indian grass, and prairie flowers now in their fall colors beyond the Rock River—the ancient home of the Winnebago Indians and more recently the Fox and Sauk Indians, some of whom still roamed about. Soon the Forbeses caught sight of the Pecatonica River lined with a broad stand of timber on its northern side. Small creeks wound their way to the river every mile or so, and just to the east of Silver Creek they reached their claim and became one of the earliest families settling in northern Illinois. Their land straddled the state road some 8 km (5 mi.) east of the settlement of Freeport, where, with axe and auger, Isaac put up a two-room log house. He chinked it with twigs and homemade mortar, spread a roof and floor of riven oak boards, and erected a great fireplace at one end. Although the house had no windows, adequate light was obtained from the fire, oil lamps, and open doors in warm weather.

During their first winter Stephenson County was established by an act of the state of Illinois, and in the spring of 1837 Isaac Forbes was appointed one of three new county commissioners. It was not, however, the best of times. After the economic downturn of 1837, money was scarce, prices low, and barter was common. Everyone had to pitch in, even the children. Flavilla Forbes, now 13, put to good use her training from Cortlandville Female Academy in New York and occasionally taught school. Grandmother Anne Sawyer Forbes wrote Flavilla from Aux Plaines in the spring of 1837:

> Dear Flavilla, With the help of two pair of glasses I sit down to try to write a few lines to you. . . . tell the children they must all be good. work well and learn well. tell Henny if he learns well Grandma will send him a new geography. I hope Francis will help his father a great deal, and Mary make all his shirts and stockings, and Flavilla teach them all she can, and their Mother Govern and Counsel them all in the right way, and so be a happy family. do

all you can to bring up your children well. I must bid you goodby.[2] (punctuation added)

Despite the hard times they held to their course. Freeport, the new county seat, flourished in part because pioneers were hospitable to new settlers. By 1840 Freeport reached a population of 491 and Stephenson County approached 3,000. Flavilla, now 16 and married, taught common school in James Hart's log cabin west of Freeport. Isaac and fellow townsmen in Silver Creek Township fenced their land, plowed the prairie sod, and planted wheat, corn, oats, melons, potatoes, and beans. Satisfactorily settled in, Isaac finally purchased the 320 acres of his squatted claim for $400 on May 24, 1842, equally divided between his farm on the state road, and his timberland across the Pecatonica in Lancaster Township.

The pioneers picked up pace as better times returned after 1843, and waves of migration rolled across the prairie. During the next decade the cast iron plow was introduced by New Englanders, a nascent copper and iron industry emerged, new canals, roads, and railroads brought people west and goods to market, and the U.S. population increased by more than 36 percent. In short, America began to take on a new form.

On May 29, 1844, Agnes Forbes gave birth to their youngest son, Stephen Alfred. Stephen's ancestors were chiefly Scottish and Dutch. There is an early paternal record of one Daniel Forbes (or Forbush) marrying Rebecca Perriman in 1660 at Cambridge, Massachusetts. Three of Stephen's great grandfathers were Revolutionary War soldiers: Stephen Forbes of Hardwick, Massachusetts, who was granted land in Wilmington, Vermont, for his army service and died there; Capt. Isaac Sawyer, a former Yale student who served in General John Sullivan's 1779 campaign in New York against the Iroquois; and Gerritt Van Hoesen, a former sergeant in a sharpshooter company under General Philip Schuyler. The latter two relatives tied Stephen's family together in a singular way: After Isaac Sawyer's death, his daughter Anne (Stephen's paternal grandmother) was raised as a young girl in the family of her father's old wartime comrade, Gerritt Van Hoesen of Cocksackie, New York. It came to pass that Anne Sawyer's stepbrother, Francis Van Hoesen, later fathered a daughter, Agnes, who in 1824 married Anne's son Isaac Sawyer Forbes.

Anne Sawyer Forbes was a remarkable woman. Unusually keen and capable, strong on learning, she was considered by the Forbes family "to be the ancestral source of whatever intellectual abilities" Stephen and his siblings had inherited. According to Stephen, his mother Agnes was "rather high strung, sensitive, [and] a devoted mother." Those who knew him well said Stephen's father, Isaac, was a "peacemaker, greatly reserved, with a most generous and sensitive nature." Stephen described him as "ironbound, kindly, generous, quick, and faithful, with ordinary abilities." The Forbes family was close and remained so, as the letters among them testify. Their "little family circle," as they put it, provided support and solace to them through the vicissitudes of their lives.

After endorsing a note for a friend to purchase land who subsequently defaulted, Isaac was left to pay the note, and thus came into ownership of 80 acres of land in December 1844, located east of Silver Creek in Ridott Township. Needing cash, Isaac then sold the central portion of his original farm and one-third of his timberland to brother Stephen V. R. Forbes, and two adjacent portions of his farm to a German immigrant. In time Isaac sold off 20 acres of his new Ridott land, and in 1848 he moved his family from Silver Creek to Ridott. At this time Stephen's oldest brother, Francis, drowned in the Pecatonica River after finishing a three-month term of courses at Rush Medical College in Chicago. The family loss was partially filled by the earlier arrival of Stephen's younger and favorite sister, Agnes Vernette (Nettie), in 1845.

And so they began once again, this time in a plain, single-room, dirt-floor cabin built of log slabs, "in which was a cook-stove at one end," Stephen later wrote, "two beds at the other, a trundle bed under one of them, a dining table in the middle, and at one side of this building a little 'lean-to' containing a single bed."

The prairie was a fine yet formidable place for Stephen to spend his childhood. Molded by wind, fire, heat, and cold, the deep-rooted sea of grasses sported a paintbox of pink, yellow, white, blue, purple, and flame-colored flowers from spring to fall. Overhead the nighthawks whizzed, grasshoppers buzzed, and thunderstorms split the sky. Young Stephen—or Bub, as he was first called—heard the drumming of the prairie chickens, learned his mother's songs from *Lallah Rook,* and clanged together musical horseshoes hung from the slab ceiling by his brother Henry. Outdoors he chased meadow mice and crickets in the grassy sward, broke open granaries of the harvester ants, eagerly watched for birds, and caught frogs in sloughs of the Pecatonica with Henry. In the hum of long summer afternoons he helped his mother pick wildflowers, and on special days, he hunted for banks of violets with Nettie. Surely the prairie was his playground.

One late fall day in the early 1850s after the ground froze, Stephen's father drove him in their wagon to the old stone schoolhouse northeast of the farm. He loved it from the start: lessons in writing and arithmetic; readings in morality, history, biography, and travel; works of Cooper, Irving, and Poe; discussions on the goodness of man and on how the strength of America lay in the nation's youth, in education, and in opportunity. These were sentimental years filled with unabashed patriotism:

God keep the fairest, noblest land that lies beneath the sun;
Our country, our whole country, and our country ever one![3]

Meanwhile, Henry taught school and furthered his own studies in Freeport and Rockford while helping on the farm when needed. Agnes and Nettie made a lengthy visit back to Preble, requiring Stephen to stay with his sister Flavilla in Silver Creek, since his father worked periodically away on the railroad. And so the Forbes farm was rented out; the family did not get back together at home until December 1852, when Stephen was eight. On return, Agnes's health and morale were poor. In letters to Henry she lamented that although "the children are happy and contented . . . I am lonely and disconsolate . . . and I am constantly confined in this dreary and desolate place." With time her health and spirits recovered, assisted by visits to and from friends and family, and by the company of her growing grandchildren, offspring of daughters Flavilla and Mary.

Clearly, however, the Forbes's fortune was at a low point. Through mortgaging and his railroad earnings, in early 1854 Isaac purchased 82 additional acres of land adjoining his farm on the north for $3.50 per acre. That summer he caught pneumonia, which quickly worsened. Expecting the end he sold and transferred possession of the farm to Henry just three days before he died at 54. "His quiet and unostentatious life," read his obituary, "was crowned by a peaceful and most satisfying death." Stephen later described Henry's response: "Without an hour's hesitation he gave up his personal plans, abandoned his career for which he had already given brilliant promise, and took upon his own shoulders the burden of our support and education; and from that time on for the next seven years he was not our guardian merely, but he was our guardian angel."

At his grieving mother's request, Henry at 21 was now head of the family. He first built a comfortable new house for Agnes, Stephen, 10, and Nettie, 9, and then took firm charge of the children's education. For both of them this meant common school until 14, as well as institutes (debates and delivered speeches), singing school, and circuit preaching, all held in the same old stone schoolhouse. Stephen then furthered his studies in literature, history, and French under Henry's tutelage at home. While continuing to help on the farm, he found the necessary time for youthful fun in sleigh rides, nutting parties, and games.

Stephen's introduction to public life occurred three months after his fourteenth birthday, during the Lincoln-Douglas debate in Freeport on August 27,

1858. Democrat Stephen Douglas was finishing up a second term as U.S. senator. Abraham Lincoln, Republican, had served earlier in the House but was defeated for a second term. Although the day was unseasonably cold with a threat of rain, several thousand people gathered in a grove near the Brewster Hotel, where a low platform had been erected for speakers and dignitaries. Lincoln arrived in a Conestoga wagon, and Senator Douglas walked from the train station. As the band played "Hail! Columbia" and "Oh! Susannah" and people milled about, Stephen pushed his way to the front of the crowd near the speakers' platform and waited with rising expectation.

Lincoln spoke first, impressing Stephen "as a truly lofty character" with his clear, calm, penetrating voice, and his "great mind in vigorous action," giving lie to Lincoln's awkward and plain appearance. In contrast, the short stocky Douglas grew more aggressive as he warmed up, eventually irritating his opponents by calling them "black Republicans," the Democrats' name for strong opponents of slavery in the new Republican party. Loud replies of "white, white" quickly came from all directions, the clamor growing louder whenever Douglas made the offensive remark. Douglas annoyingly reminded his audience that while Lincoln spoke Democrats did not interrupt *him*. Not able to contain himself, Stephen shouted out that maybe that was because "Lincoln didn't use any such talk!" Adults nearby scolded Stephen that he must not "talk back." Taking the platform again, Lincoln vindicated Stephen by commenting that the Democrats treated him (Lincoln) like a gentleman, because he treated them as a gentleman should, bringing hearty laughter and vigorous applause from the audience.

Much later, Forbes wrote that he had read Harriet Beecher Stowe's *Uncle Tom's Cabin* in the early 1850s, and that although he was perhaps predisposed to Lincoln's views, he had no conscious party preference before the Lincoln-Douglas debates, but that after reading her book all had changed. "I came away from [the debate] quite aflame with enthusiasm for the new Republican party and especially for Lincoln as its champion. . . . these boyish impressions fixed my politics, as it proved, for life, and did more than anything else to send me into the Union Army only three years later and to hold me there until the end of the Civil War."[4]

Recognizing Stephen's quick mind, enthusiasm for learning, and omnivorous reading habits, Henry arranged for Stephen at age 15 to matriculate for the three-month 1860 winter term in the preparatory department at Beloit College, across the Wisconsin border some 50 km (31 mi.) northeast of Ridott. (They took the train together from Freeport in late December 1859, but Henry walked back home for lack of ticket money.) One hundred and eight boys from nine states were preparing for the regular college course, each paying $20 tuition. Stephen enrolled in middle prep, taking rhetoric, classics, and philosophy, with Greek as an extra subject.

"I miss you from our family circle," wrote his mother in early January, "but content myself thinking that it is for your benefit that you are absent." Agnes later wrote that she was glad to learn he was "getting along well" in his studies and worried that he might be "somewhat homesick." For his part Henry was concerned with Stephen's expenses, urging his brother to "keep an eye out" for cheaper boarding but not to let this hinder his studies. Initially Stephen spent $2.50 per week for board, but he eventually made other arrangements. He continued to study "like fury" in full appreciation of Henry's occasional stipend of five dollars and his family's slim resources.

Stephen's term at Beloit stimulated and strengthened his early interests in language, literature, and philosophy, interests that he would soon strive valiantly to cultivate and conserve under conditions of horrific deprivation and extreme danger. A favorite book of his at the time was Jonathan Edwards's *Freedom of Will* (1754), in which the theologian vigorously refuted self-determinism, arguing that an act of will does not come into existence without a cause and that "there can be no virtuous choice unless God immediately gives it." Two short essays that Stephen wrote at Beloit that winter reflected Edwards's influence. In "The Dignity of Reason," Stephen argued that divine revelation provides a check on mankind's imperfect reasoning, hence "if perfect Reason keeps us from all error, the more we cultivate and develop our reasoning faculties, the less liable to sin shall we be." And in "Do Noble Things," Stephen wrote that although "there is a natural longing in the soul of every person for high and noble deeds . . . still what merit there is [in a noble act] belongs not to himself or his will . . . but to his nature, over which he has no power."

Soon the winter term drew to a close. Henry wrote, "I would come up to your examination, but I can't spare the dollar forty." More important, Henry informed his brother that he could not raise the money for Stephen to attend spring term at Beloit but would instead send him for two terms the following academic year. In the meantime he supposed, "You will doubtless be able to help yourself by teaching."

The weather was magnificent that spring—warm, sunny, and dependable, hence the wheat and oats went in early—and Stephen walked comfortably home from Beloit after pushing his trunk to the railroad station by wheelbarrow. Still wound up after his intensive academic work, Stephen threw himself into frenetic activity with friends and work on the farm, eventually paying the price of fatigue followed by several weeks of forced inactivity under his mother's care in the family circle.

Years afterward he recalled how the four of them—Stephen, his mother, Nettie, and Henry—sat in the dark together on Sunday evenings in their prairie cabin and talked and talked: "We felt as if we were under some wonderful, but winning enchantment, as if the spell of silence and night had isolated us from all the world and from our individual interests, and it was no longer mouth speaking to ear, but soul to soul."

Then May gave way to June, the blackbirds, meadowlarks, and snipes returned as haying got underway, and news came that Abraham Lincoln and Stephen Douglas would debate each other once again, but this time as their parties' nominees for president of the United States. After helping with the summer wheat harvest, early in September Stephen left for a Methodist camp meeting across the Mississippi River in Sabula, Iowa, where he stayed with his sister Mary Foster and her family. On his return to Ridott, Stephen continued "studying ahead on the college course" with Henry's assistance and working on the farm. It would be only a matter of months before events would take place that would severely test the country and change the course of Stephen's life.[5]

CHAPTER 2

The War Years, 1861–65

The opening of the Civil War in April 1861 prompted heated discussion in the Forbes's family and led, that summer, to Stephen and Henry's enlistment at Cedarville, Illinois. They called themselves "Winnesheaks" after a Winnebago chief and village of that name from the Freeport area. Forbes later remarked on their state of mind:

> To remain quietly at home, busy with books and teachers while my comrades were thronging to the front to fight for *my country*, was simply impossible, and the mere thought of it intolerable. Indeed, I think, that most of us were secretly glad that we had been born in a time when it was possible for a boy to do anything so wildly and gloriously different from what had been planned for him as to go to war. It was not to us a dilemma, a sacrifice; it was a privilege, an intoxicating opportunity; we could not be made to stay at home.

Henry was commissioned as a first lieutenant and married Jennie Gorham in September before leaving for camp. Stephen Forbes, a private at 17, wrote later that he and Henry "were wholly without special preparation for war," but believed that "the life of the nation was in mortal danger."

They arrived at Camp Butler near Springfield, Illinois, during the third week in September 1861. "We are here in tolerably good health and extremely good spirits," Forbes wrote his sister Nettie. Camp life agreed with him, and he liked it "better and better" as he grew more accustomed to it. "Our hardships here are not nearly as great as one might imagine," and he would not "change with you for anything I can think of." The lifestyle certainly suited him.

The longer they were in camp the more he and his comrades found it "most especially agreeable": up at 5 A.M., regimental band playing "Hail Columbia"; the long line of white tents; sentries with musket barrels gleaming; breakfast of bread, meat, and coffee; foot drill; pitching quoits; cracking nuts; dinner; horseback drill; supper; the campfire where they wrote, told stories, strolled, or prayed; 9 P.M. tattoo and bed; and occasionally guard duty, thinking "thoughts of the home behind and the battlefield ahead." And so their days went.

In mid-October the Forbes brothers' Seventh Illinois Volunteer Cavalry Regiment, Colonel William Kellogg commanding, was mustered into U.S. service. Soon after, Henry was chosen as captain of Stephen's Company B; and as for Stephen, he said he was "getting as tough as a young bear."[1]

Marching orders came for the regiment in November and early December 1861. They struck their tents and left eventually by boat for Birds Point, Missouri, across from Cairo, Illinois, at the confluence of the Mississippi and Ohio Rivers. Confederate troops and sympathetic guerrilla bands were a threat there to the Union right flank on the Mississippi. They were now in enemy country. Forbes later said that this was "where I made my first scout and saw my first dead soldiers." Forbes would spend that winter scouting and skirmishing, mounted and armed with his saber and a Sharps carbine, both appropriate for cavalry patrols through timber and swamps.

The western Union army's primary mission in late 1861 was to open up the Mississippi River and loosen the Confederate grasp from Cairo, Illinois, to Memphis, Tennessee. Union General John Pope would advance his army of 25,000 men on New Madrid, Missouri, and the Confederate stronghold, Island #10, on the Tennessee side (fig. 1).

Forbes's regiment set forth in early March 1862, to Riddle's Point, below New Madrid to protect Union batteries and keep an eye on a half-dozen Confederate gunboats and batteries on the wooded eastern shore. New Madrid fell on March 14 to General Pope's bombardment. Now he had to get his transports and gunboats downstream past heavily fortified Island #10. He did this via a canal dug by his engineers through the swampy peninsula to the east of New Madrid. Pope's shallow-draft transports then floated into New Madrid and out again into the river bypassing the island.

Forbes reported on cannon fights, "booming of guns at No. 10," and driving back "five rebel gunboats." Union gunboats *Carondolet* and *Pittsburgh* had also silenced a number of batteries on their way downriver. General Pope soon brought his army across river in early April, forcing the surrender of 7,000 Confederates on the island and in adjacent garrisons. The Mississippi was now open to Memphis.

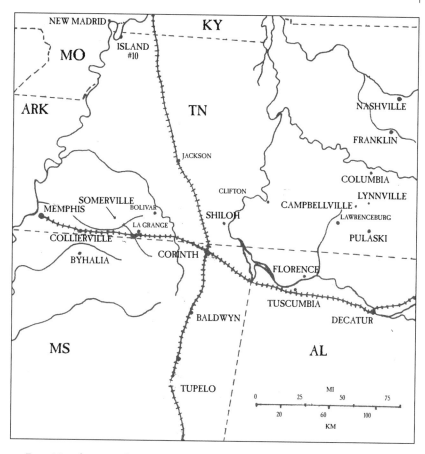

Fig. 1. Main theater of military operations of the Seventh Illinois Cavalry Regiment, U.S. Volunteers, 1862–65 (except for Grierson's Raid).

That April, as the weather warmed, Forbes found the surrounding Missouri landscape very pleasing to his eye. "We might almost forget that we were here with such warlike intent," he wrote, "and imagine that we were just out pleasuring." The nascent spring woodland along the river moved him deeply as he described it to Nettie while recalling their childhood rambles: "The surrounding forests ringing with the songs of birds are adorned with scarlet masses of redbud. The low moist ground . . . is fairly spattered with buttercups, looking as if one of childhood's extravagant wishes was realized and the fleecy clouds had rained a shower of gold, while thick clusters of violets spangle the mold and leaves with patches of blue."

Plainly, the rolling Mississippi had become a familiar friend to Forbes and his comrades, evoking memories of home. Its surface reflected the moonlight and stars, its running ice made winter music, and its mist and mud embraced them. But now their time was up as they received marching orders for Tennessee.[2]

Forbes's regiment left New Madrid on the steamer *Estella* on April 15, 1862, sailing up the Mississippi, the Ohio, and finally the Tennessee River, where they disembarked near Shiloh, site of a recent battle between the Union and Confederate forces. Overall commander of Union forces, General Henry Halleck, would soon have a giant army of 125,000 men facing a Confederate force of about 60,000 under General Pierre Beauregard at the important railroad junction of Corinth, southwest of Shiloh in the northeast corner of Mississippi.

Company B skirmished several weeks with Confederate units on roads around the small town of Farmington just east of Corinth. In early May, Major Aplington of their regiment was killed. Henry then led Company B, with Stephen newly promoted to sergeant, four miles to a river ford. Moving through thick woods in stifling heat, Henry spread his flankers, and his advance guard soon made contact with the enemy. Galloping horses in front raised a chorus of rebel yells, a sharp volley of firing, and the quick reply of smoking Union carbines along the battle line.

Henry shouted, "Stand firm! Don't flinch! Hold your ground men!" The enemy divided and swirled left and right as several troopers were struck with Union fire and fell from their horses dead.

Forbes reflected to Nettie on what he had just seen: "I used to think that after one had killed a man he must be altogether a different being, that a change would come over him like the taint of leprosy when he realized the burden of death that was upon his soul, but it is not so, there is no [such] change in the man."

Late in May, General Beauregard, outnumbered 2 to 1, retired south with prearranged deception. To learn his movements, Union General Halleck ordered a cavalry probe south of Corinth. And thus it was that Sergeant Stephen Forbes rode out early in the morning of June 2, 1862, carrying a dispatch to a Union cavalry outpost. Galloping down a winding road, Forbes was suddenly surrounded by enemy soldiers with no chance of escape.

Forbes told his captors he had only a verbal message, so he was threatened with hanging and would now have to answer to General Braxton Bragg at his headquarters nearby. Fortunately, Forbes managed to tear up his dispatch and stuff it into his pistol holster. Bragg had Forbes searched and sent him to higher headquarters at Baldwyn. There Forbes was questioned by a venomous major and then, to his amazement, by General Beauregard himself. Forbes was then taken in a guardhouse occupied, "by a dozen or so friendly Union men who welcomed me as if we were dear friends meeting after a long separation." So began four and a half months as a Confederate prisoner of war. The story that follows was selectively composed from Forbes own prison journal.[3]

After a hearty supper of biscuit, molasses, and fried bacon, Forbes and his fellow prisoners traded tales with high spirits and comradery. For the next few days they taunted a group of Confederate deserters who shared their prison. On June 7 they were told to get their bundles together and were marched to a train that rattled south 55 km (34 mi.) to Okolona and beyond into "the finest agricultural region I ever saw." Then onto Meridian, Winchester, and finally on June 9, finishing their 475-km (295-mi.) journey to Mobile, Alabama, where the Confederate deserters left them.

Marching under guard through dusty streets, their sorry column passed cotton houses, slave pens, and groups of gawking, curious citizens. Soon they reached their appointed prison surrounded by a 20-foot-high brick wall.

There were fifteen of them now, mostly Second Iowa calvary troopers. The food was good but the water poor. Forbes was surprised to find a small assortment of books. To add to it, Forbes walked into town with an armed guard once to buy a bible and another time a Greek grammar he hoped would "turn my idle time to good account." Mobile appeared as sullen and depressing as his prison room, so he was delighted to pluck an almond flower on one of his walks that he put in a small bottle of water, cheering up his room and spirits too.

As June wore on their idleness told: "A mountain of dreary ennui sits upon our spirits," while thoughts of freedom filled their endless hours. In mid-June the Confederate provost marshal told them they would be sent to Macon, Georgia, and be "speedily exchanged," but dare they believe it?

On July 4 their group, which had now grown to thirty, boarded an ancient steamboat, the *Senator*, and ended up in the stinking engine room along with some slaves, a few chickens, a pig, a baby alligator tied in a tub, and an army of fleas in their sleeping sacks. Steaming up Mobile bay, they approached the marshes lying between the Alabama and Mobile Rivers, whose waving grass and sparkling water Forbes had viewed from his prison room. Nibbling on dry bread and almost putrid ham, they anticipated every curve of the Alabama River while guards with bayonets watched over them.

Passing through Selma and Montgomery, Alabama, they then left the steamer, and on July 9 boarded a train with twenty more prisoners bound for Macon, where they marched with a Confederate sergeant to a prison yard marked "Camp Oglethorpe." The 10-acre yard had wooden sheds crouching on dusty ground near stands of oak and pine, and nearby, some well-worn paths the prisoners paced for exercise. On the south side lay a forest, its canopy providing breezy comfort in the cool of the day, its little brook coursing through the yard—breaking their hearts and moving on. Sorrow now settled over them like a shroud: most had scurvy, chances of exchange had evaporated, escape seemed hopeless, and they had little to do. Yet, Forbes had been planning an escape lacking only a partner. And once again rumors of exchange flew among them just as Forbes had about decided to escape alone.

For several days Forbes painfully watched the delirious struggles of his dying friend Henry Wilson, a sergeant of his regiment, his "first experience with a desperately sick man." Despite Forbes's nursing care, one night he found Wilson dead of typhoid fever, far from Illinois, lying under a canvas sheet in the bright moonlight. Memories of Forbes's Illinois prairie home now surfaced:

> Fields of waving wheat and rustling corn clothe the old homestead . . .; the larks and bobolinks are singing to-day by the old spring in the meadow; the thrush is pouring his notes into the clear morning air from the broad-boughed apple-tree; the roses are blossoming . . . by the roadside; and the billowy fields of grass are waving in the sunlight. . . .
>
> I am sick to death of the horrible degradation and bestiality of this place, where . . . no vileness is too low to be practiced. . . . My whole soul goes out with a longing cry for the green fields of home."

Early in August, Forbes had his first malarial attack. During the next two months with alternate chills, shaking, and burning, he grew too sick to walk, crawl, or even care as his small supply of quinine ran out. Quickly he sunk into deep depression over the struggle against "vice, degradation, and filth, moral, mental, and physical." He found himself turning away friends and repulsing strangers, berating himself, and then withdrawing. He had no fear of dying. What he feared most was "being gradually overcome and infected by the horror and misery of this place." Slowly he realized that these prisoners needed help and comfort, help to connect them "with the memories of our former life." He himself had benefited from the kindness of several of them. Now he seemed reminded that these prisoners, although not in his regiment, were his comrades too, that they needed each other; and it was in everyone's best interest that they strive to be humane.

Forbes had discovered that war evoked both competitive and cooperative human instincts. The fates of the prisoners were linked. Cooperative behavior would nourish respect for each other and especially tolerance. Tolerance in the widest sense was a trait that Forbes would exhibit increasingly through his adult years and one he would employ in both his private and professional pursuits to the inspiration of all.

In this fresh frame of mind Forbes met nine men from his regiment in a group of forty new prisoners arriving from northern Alabama. They informed him that his brother Henry was in good health and still commanding Company B.

Early in October 1862 they left Camp Oglethorpe destined for Richmond, Virginia. Seven hundred men received rations and then boarded twenty-odd boxcars with rough plank seats. The next morning the train reached Columbia, South Carolina, where once again they marched through city streets barefoot and ragged while they gazed awestruck at beautiful gardens and homes. Fully one-third of them could hardly walk, and several men died along the roadside.

Now the specter of death stayed with them constantly: in North Carolina, where Forbes gave sicker men some of his clothes, and men lay dead and dying on the floor of an engine house and in an open lot where five more comrades died by morning. The living were perhaps saved when concerned citizens of Raleigh, North Carolina, sent meat and soup. The next day (October 14) they left Raleigh in open cars and were showered with apples by children and slaves along the way to Petersburg, Virginia. To complete their 2,000-km (1,240-mi.) journey, they were finally deposited in one of a half-dozen old tobacco houses used as a prison in Richmond.

Forbes was exchanged in mid-October 1862. He steamed to Washington, D.C., on the Union ship *Commodore*, where he luxuriated in freedom on the Potomac River during four exquisite Indian summer days. Soon the *Commodore* embarked on the 960-km (595-mi.) sea voyage to Rhode Island, arriving in Narragansett Bay late in October with a seasick and lonely young soldier aboard. Forbes was admitted to the U.S. General Hospital, Portsmouth Grove, Rhode Island, as a convalescent paroled prisoner with malaria and scurvy on November 2, 1862, five months after his capture. He later remarked on the effect of his prison experience: "It was [in prison] that some of us youngsters "found ourselves," and it was here . . . that the green, reserved, silent, introspective, and unobservant country boy became a man, and here that he learned to understand and appreciate other men until that time a riddle to him."[4]

He was discharged from the hospital on December 19, ordered to St. Louis, and returned to his regiment in La Grange, Tennessee, in early January 1863.

Forbes's regiment, in addition to the Sixth Illinois and Second Iowa cavalry regiments, were brigaded together now under Colonel Benjamin Grierson at La Grange, midway between Corinth and Memphis in western Tennessee. Forbes was now first sergeant of Henry's Company B. That winter and spring their feelings ranged from depression to quiet confidence in action to boredom during periods of inactivity. Forbes seemed to accept his own role as necessary, but the future was "dark and uncertain." He wrote Nettie, "the way things are shaping, I have about made up my mind that I shall have to die in the army," and better on the battlefield than in prison, he felt. He prayed "to keep this horror of [war] away from you at home."

Early that year the Confederacy still controlled some 400 km (248 mi.) of the Mississippi River between strong fortifications at Vicksburg and Port Hudson, the latter located just north of Union-held Baton Rouge (fig. 2). To capture and control the Mississippi valley and win the war, this section of the river had to be taken. The basic Union plan was to attack at both ends: General Grant above at Vicksburg and General Nathaniel Banks below at Port Hudson. Defending this region for the Confederacy was General John Pemberton with 38,000 men under his command and headquarters at Jackson.

The mission of Colonel Grierson's brigade, including the Seventh Illinois Cavalry Regiment, was to strike deep into Mississippi and destroy the railroads in the rear of Vicksburg between Jackson and Meridian, cutting wires, burning provisions, and "doing all the mischief you can." Their raid would serve as a strong diversion, attracting Confederate attention and troops while General Grant planned and initiated his attack on Vicksburg.

Well-rested after a long winter and eager for action, Grierson's Union troopers, Forbes wrote, "each carried 40 rounds of ammunition, five days rations, and a good supply of salt." At sunrise on April 17, 1863, Grierson's brigade of 1,500 men now darted from La Grange and moved south. The brigade included Battery K, First Illinois Artillery, with six two-pounder guns. Three to five days out, between New Albany and Starkville (see fig. 2), Grierson slyly sent out smaller detachments as decoys. One of these drew the attention of Lieutenant Colonel Clark Barteau, commanding all Confederate cavalry in northeast Mississippi. Soon after General Pemberton learned of a Union cavalry raid in that area and requested more Confederate cavalry to be sent that way. At that moment, however, Lieutenant Colonel Barteau's cavalry was the only force Grierson had to contend with.

Grierson quickly detached Colonel Edward Hatch's Second Iowa Cavalry on April 21 with 500 men to confront Barteau's force. They skirmished as Hatch drew Barteau steadily northward and eventually returned to La Grange with a loss of only ten men. Grierson, left now with the two Illinois regiments of 950 men, sped southward with few known organized enemy forces between him and the Vicksburg Railroad.

On April 22 Grierson sent Captain Henry Forbes and thirty-five men of Company B, including Stephen, out on his left flank. Their orders were to approach Macon, Mississippi, cut wires and track, and create confusion. Captain Forbes's small company did well, giving the enemy the impression that a Union force of several thousand was involved. Rapid arrival of larger Confederate forces, however, precluded any further action from Captain Forbes, who

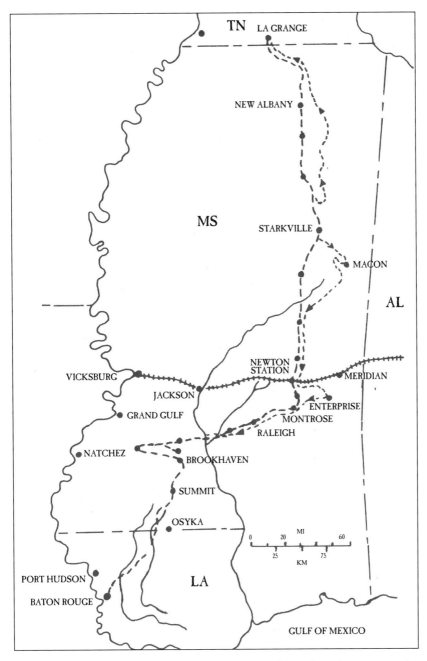

Fig. 2. Grierson's Raid, April 17–May 5, 1863. The heavy dotted line indicates the movement south of Colonel Grierson's brigade. The lighter dotted lines with arrows show the movement of Colonel Hatch's Second Iowa Regiment and Captain Henry Forbes's small company from the Seventh Illinois Regiment. (Adapted from a map in Scott [1936], Forbes Family Letters, *204.)*

next day rode south toward a hoped-for reunion with Grierson. So far, Stephen Forbes wrote, "this gallant little party" had already "accomplished quite as much as if it had been ten times as large." Yet they would do even more.

Meanwhile, Grierson had galloped from Louisville on April 22 to Decatur on April 24, when he finally struck the Vicksburg Railroad at Newton Station at 6 A.M. Forbes later wrote: "Two trains of [20] cars were captured and destroyed, one filled with food and ammunition, including several thousand loaded shells, and the other with machinery and railroad ties. Commissary and quartermaster's stores were burned, five hundred stands of arms were broken up, seventy-five prisoners were captured and paroled, and the railroad was wrecked and its bridges were burned . . . for four and a half miles to the east."

At this point confusion ran rampant at General Pemberton's headquarters as multiple orders went out and complex movements of troops were made. Grierson had proved so elusive that a baffled Pemberton said that Grierson's cavalry had been reported "in various places at the same time." Colonel John C. Taylor of Pemberton's staff wrote in his diary that this raid had caused a "thundering rumpus," drawing attention away from Grant's threat across the Mississippi. Of immediate importance, Pemberton feared that his supply line to Vicksburg was now in jeopardy and that even Jackson, the state capitol, was endangered. The only advantage he had was that Grierson had now lost the element of surprise.

The Forbes brothers' Company B in the meantime searched in vain for Grierson. They did find a company of Confederate home guards with which they fought, capturing 230 and fresh horses. Soon after one of their troopers was killed and another wounded by Confederate stragglers. Aware that their presence was known, they now rode "watchful of every turn in the road." At dawn on April 25, they reached Newton Station, whose ruins still smoked fifteen hours after Colonel Grierson's command had passed south.

Just about that time Grierson breakfasted near Raleigh in Piney Woods country and contemplated his next move. Learning that a battle was expected near Grand Gulf, Mississippi, 160 km (99 mi.) to the west, he decided to move that way and attempt to join General Grant in his Vicksburg campaign. This would carry his brigade southwest to Raleigh and Westville, then across the Strong and Pearl River bridges and into Hazelhurst on the Southern (Jackson) Railroad by the twenty-seventh, deeper and deeper into Mississippi country. He was now burning all bridges as he crossed them, implying he had little hope for the return of Company B anytime soon.

In the afternoon of April 25, 1862, General Pemberton in Jackson received a message from General William Loring at Enterprise, Mississippi, saying that a Union force "represented as 1500 strong" had demanded surrender of the town. This reported force, however, was Captain Henry Forbes and his small company of thirty-four men. After leaving Newton Station that morning, Captain Forbes believed that Colonel Grierson had probably crossed the Mobile & Ohio Railroad at Enterprise and rode on into Alabama for the return trip north, since this was the tentative plan he originally made. The company reached Enterprise at 1 P.M., where they were greeted with shots from a stockade. Captain Forbes tied a truce handkerchief to his saber, which promptly produced three Confederate officers riding out with their own truce flag.

"I come from General Grierson to demand the surrender of Enterprise," said Forbes, promoting Grierson for effect.

"Will the Captain await the reply here?" asked one Confederate officer.

"We shall fall back to the main body," replied Forbes, "and there await the reply."

Not receiving a reply after an hour wait, Captain Forbes and his men rode rapidly west toward Garlandville (near Montrose), where some of his disguised scouts found that Grierson had indeed passed this way to Montrose and westward. After resting they galloped to Raleigh and Westville. Captain Forbes now estimated that Grierson's column was only 7–8 hours ahead. They needed to reach him before he burned the Strong and Pearl River bridges, since neither river was fordable without them.

Three men volunteered to ride rapidly ahead at 5 A.M.: Privates John Moulding and Arthur Woods and First Sergeant Stephen Forbes. "I never expected to see one of them again," Henry Forbes later wrote, "feeling sure that they would be picked off by stragglers." Just after sundown in a drizzling rain, they caught the sounds of a marching column ahead. They approached Grierson's rear guard, who challenged and then waved them on. Forbes rode up immediately with a message to Grierson at Strong's Bridge: "Captain Forbes presents his compliments and begs to be allowed to burn his bridges for himself," he rasped.

Using a single flatboat carrying twenty-four men and horses at a time, the brigade then crossed the Pearl River by the night of April 27 after convincing the ferryman they were men of the First Alabama Cavalry. (By now the raiders had donned captured clothing.) Elements of the brigade then proceeded to destroy track, trains, wires, stores, track, and ammunition north of Brookhaven.

Not wishing to be a sitting target and having received no word of Grant's crossing the Mississippi, Grierson decided not to march toward Grand Gulf, but instead to move rapidly south to Baton Rouge. Next morning, April 29, after sending the Sixth Illinois Cavalry in a strong feint westward out the Natchez Road, Grierson suddenly countermarched the brigade via back roads and, by dark, had passed south beyond Brookhaven on the New Orleans & Jackson Railroad. They then saw to it that "the railroad from Brookhaven to Summit was as badly wrecked . . . as any road could well be in so short a time." Forbes himself was detailed to burn the Brookhaven Station while the brigade also destroyed cars, tracks, wires, trestles, depots, and a small bridge.

Events during the past seventy-two hours left General Pemberton groping for Grierson's intended course. In a creeping state of uncertainty, Pemberton and his overwhelmed staff tried to guard all possible points as southern Mississippi swarmed with Confederate cavalry. On the night of April 30, Grierson and his men camped 19 km (12 mi.) below Summit in the forested watershed between the Amite and Tickfaw Rivers.

Inevitably some Confederate officer and his command would get in advance of Grierson's line of march. Major James De Baun with his Ninth Louisiana Partisan Rangers had come up from Port Hudson toward Osyka, reaching Walls Bridge over the Tickfaw River midday on May 1. Resting there his foragers warned him of Grierson's approach on the Summit road, allowing him time to conceal his 115-man battalion in an ambush below the bridge. In the ensuing skirmish two of Grierson's companies charged dismounted, while two cannon made matters hot for Confederates in the woods. De Baun's men eventually scattered toward Osyka, leaving two raiders dead and three combatants wounded, including Union Lieutenant Colonel Blackburn mortally, and a half-dozen raiders and rebels alike taken prisoner.

And now on the fifteenth day of the raid, they faced the longest, fastest ride of all. They were then approximately 100 km (62 mi.) from Baton Rouge, but of immediate importance for them was to reach the Williams Bridge spanning the unfordable Amite River east of Port Hudson, commanded by Confederate General Franklin Gardner. Shortly they reached the Louisiana line north of Greensburg, and later in the evening they heard the boom of big guns off to their right as federal mortarboats shelled Port Hudson. At midnight their advance guard swept down to the Amite Bridge, capturing two unsuspecting guards, "and in a few minutes," Forbes later reported, "just as the moon rose to light our way, the muffled thunder of our horses' feet resounded from its entire length," as the troopers sped on.

Despite being hungry and tired, they could not stop, still having 35–40 km (22–25 mi.) to go. Many men rode sound asleep and some strayed, providing the Seventh Illinois—as rear guard—with all the work they could handle in urging their comrades on. Meanwhile the advance guard captured two Confederate camps holding 80 men: one at Sandy Creek and the other at a ford of the Comite River near Baton Rouge. Since leaving La Grange on April 17 they had now killed or wounded 100 enemy and captured more than 500 prisoners, with most of the latter paroled (until exchanged for Union men). Now the guard switched, with the Seventh Illinois marching in front during the early morning hours of May 2, through bayou, woodland, and sugar plantations from which slaves spontaneously appeared, alerted no doubt through their communication network bringing news of Yankee raiders moving south.

Several hours later a Union cavalry company from Baton Rouge appeared, also having heard rumor of the arrival of "an important force." They had trouble believing the tale of this motley, tattered crowd, however, as did Major-General Christopher Augur, Union commander at Baton Rouge. In fact the Department of the Gulf Army had not even been informed by General Grant that a cavalry raid was in progress across Mississippi because Grant assumed that Grierson would return to Tennessee back through Mississippi or Alabama. A puzzled Augur met with Grierson at headquarters, and it was not until then, Forbes later wrote, "that we were admitted to the federal lines and to the protection of the flag." Once convinced, Augur insisted on a parade in honor of the brigade that afternoon, a proposition less attractive at that juncture to Grierson and his men. Nevertheless, a parade it was, and in the following order: Sixth Illinois, four 2-pounder guns and the Artillery Company, Confederate prisoners, Seventh Illinois, additional horses and mules, and 300 blacks bringing up the rear. "As we rode at last through Baton Rouge, the streets were banked for a mile or more on either side with cheering crowds of citizens of the town and the soldiers of Augur's army, and the wayworn but triumphant column was brought to bivouac in a beautiful magnolia grove to the south of the city."

For his gallantry and leadership, Henry Forbes was promoted to major. General Grant reported that Grierson "has knocked the heart out of the state. . . . It has been one of the most brilliant cavalry exploits of the war." By mid-May northern citizens learned of the events when newspapers across the country carried the story of this 965-km (598-mi.), first successful Union cavalry raid of the war. The accolades went far toward improving the reputation of the Union cavalry, as well as reinforcing what Grierson already knew—that his western troopers were exceptional.

Following the Grierson raid Forbes's regiment joined the First Cavalry Brigade commanded by newly promoted Brigadier General Grierson attached to General Banks's army at Port Hudson, Louisiana. Forbes fought near the fortress against some of the same Confederate cavalry his company had encountered on their recent raid. In addition Forbes went to the front twice "on his own hook" in assaults during May and June. Port Hudson finally fell to

Union forces on July 9, five days after Vicksburg fell to General Grant, giving the Union control of the Mississippi River from Memphis to the sea. Forbes and his regiment had earned a much-needed pause and change of scene.[5]

Forbes reached Memphis from Port Hudson on July 22, 1863. He had been impressed with the Louisiana Gulf country—its rich bottomlands, groves of magnolia and acacia, dark forests, and backwater bayous—but he was nonetheless happy now to find a grand world of clean clothes, morning papers, and living in tents for the first time in more than three months. His regiment left Memphis in late August riding east to Collierville (see fig. 1) on the Memphis and Charleston Railroad. Here they would prepare for joint cavalry and infantry operations during the next six months, protecting vital lines of supply and communications.

In mid-October Forbes's regiment rode westward along a branch of the Coldwater River southeast of their Collierville camp. They were part of Colonel Jess Phillips's cavalry brigade of 1,100 men, including two other Illinois regiments and the Seventh Kansas. Colonel Edward Hatch, now cavalry division commander, went along. They sought General James Chalmer's Confederate cavalry. They found one of his 800-men brigades on hills south of Byhalia Creek. Several companies of the Seventh Illinois were in the initial charge as First Sergeant Forbes rode up with his commanding officer, Lieutenant William McCausland, and men of Company B. They formed a battle line and advanced on the enemy under heavy fire.

"Hold on boys!" cried Lieutenant McCausland as he turned in his saddle and moaned, hit by enemy fire. Forbes led the officer to a tree for safety, gathered up the company, dismounted, and charged through a cornfield up to a road with carbines blazing. Suddenly, Forbes wrote, "a great roll of artillery smoke puffed out from the woods, and the shouting grape-shot came tearing at us." Circling back with his men, Forbes checked on Lieutenant McCausland, who had been shot through a lung. After calling for a stretcher bearer, Forbes and his men joined a larger line to attack the enemy left as Union cannon shelled on their enemy's right as a diversion.

Three regiments now went forward firing as they went, tree to tree, log to log, with enemy shot and shell whistling around them. In an instant Forbes saw Lieutenant Nicholson of Company M shot through the head and killed and other Union troopers falling wounded. The enemy, however, was soon in full retreat after a three-hour fight.

Unbeknownst to Forbes but fittingly, his commission as second lieutenant had been signed by the governor of Illinois four days before. He was nineteen years old and his yearly pay would be $1,350.

A few weeks later, after Forbes had put on officer's shoulder stripes, he commanded his company for the first time in his new grade as an officer. The occasion was Confederate General Chalmer's return to destroy the railroad in the Collierville area. Forbes was on picket (outpost) duty with his company and Company M, south of Collierville, when they were surrounded by advance units of seven regiments of enemy cavalry. His force had four men killed or wounded and more than half captured. The ensuing battle of Collierville on November 3 left 95 Confederates killed, wounded, or captured, and 30 Union troopers likewise. The advantage this day lay with the Union forces and the Colt revolving rifles of the Second Iowa as they broke and repulsed an enemy who assumed they were charging single-shot carbines.

But soon Union forces in western Tennessee would face an opponent so formidable, that any advantage they gained with their arms would be almost irrelevant, and any tactic they used other than resolute attack would invite certain defeat. That opponent was General Nathan Bedford Forrest, cavalryman extraordinaire.

Forrest's mission in mid-November 1863 was to raise recruits and bring together irregular partisan units for CSA service. To do this he brought along approximately 450 men, including his legendary 65-man escort company, Mac-Donald's battalion of Kentucky cavalry, 150 men of his brother Jeffrey's regiment, and a battery of artillery—a smallish force, but in Forrest's hands a keen instrument for heroic acts and for the bafflement of many a Union commander. To them he was "that devil Forrest." To his men he was "Old Bedford," for whom they would fight and follow with their last measure of devotion. Some Confederates said they feared the general more than the enemy and others that he asked nothing of his men that he did not ask of himself. Wise soldiers knew he had the gift of making the most use of whatever force he had: "Git thar fust with the most men," he was said to have declared—or at least make the enemy believe that was so. All knew he would invariably meet a charge with a charge, concentrating his power of attack at the most precise moment. Even in the twilight of their lives the men of the Seventh Illinois would still remember the thrill and terror of Forrest's thunderous charge in the line of battle and know they had once met and fought one of the great soldiers of the Civil War.

At age 40, when he joined the Confederate army as a private, Forrest was already wealthy by way of land, livestock, cotton, and slave transactions. Despite an almost complete lack of formal schooling and no active military experience, he rose to the rank of lieutenant general. Wounded four times, he had twenty-nine horses shot from under him and personally killed thirty Union soldiers in hand-to-hand combat. Finally, he had no use for drill and parade-ground formalities nor did he know much of the art and theory of war; instead, as he said, he just "fought by ear."[6]

On December 2 Forrest crossed into western Tennessee. Forrest then moved rapidly north to Jackson, where he raised troops and gathered supplies so successfully that two weeks later he sent a full mounted regiment safely south. Meanwhile, Union Generals Stephen Hurlburt and Grant planned multiple cavalry movements against Forrest. Both sides, however, now had to deal with awful roads, rain, sleet, and snow, and the swollen Hatchie and Wolf Rivers. Union cavalry units under Generals A. J. Smith, William Sooy Smith, Joseph Mower, George Crook, and Grierson then moved toward Forrest's expected area of advance from every compass direction.

Forrest started back south on December 23 after receiving warning of approaching Union cavalry. He pushed forward two advance detachments that skirmished Union units, crossed the Hatchie River by ferry, and attacked and drove away Forbes's Seventh Illinois near Somerville on a frosty Christmas Eve 1863.

A few days earlier Forbes had written Nettie that he hoped to tell her "of the complete thrashing of our archenemy Forrest." The Seventh Illinois had their chance again though on December 26, when they fought Forrest for three and a half hours at Somerville. Forbes wrote: "Surrounded, cut to pieces, and severely thrashed, although we fought splendidly." In Forrest's view: "Succeeded in getting in their rear and cutting them up badly, capturing their wagons, a good many of their arms and horses, and 45 prisoners, and killing and wounding 28." Even Stephen's mother joined in with comments concerning Forrest: "I hope you will destroy him and his whole army."

During the last week in December, Forrest marched west from Somerville and then south. Sending his main column over the border into Mississippi, Forrest and his best fighters stayed behind to attack Union forces arriving at the rear. In a driving and blinding rain and in what Forbes called a "spirited fight," Forrest held or drove back the Seventh Illinois and other Union units until after dark. Heavily outnumbered and with masterful cunning, Forrest had misled and outfought his opponents, returning on December 29 with prisoners, 3,000 recruits, his artillery, and wagons.[7]

During late 1863 and on into 1864, the subject of reenlistment came up in letters among the Forbes brothers and their family. Three-year terms of enlistment for many regiments mustered in during 1861 would end during the latter half of 1864, meaning a potential loss of experienced Union soldiers. Any regiment where 75 percent of those men with less than fifteen month's service

remaining would reenlist for three years or the duration of the war would retain its identity and officers and be designated a "veteran regiment."

This issue immediately brought up what postarmy plans either Forbes man
had in mind. Family expectations for education and professional futures for
those Forbes children with promise were high, especially with his mother. Although Agnes was demonstrably proud of Stephen's military "deeds and noble
exploit," she reminded him, "It is but little more than nine months before
your time is out and then I want you to finish your Colledge [sic] course and
choose your profession." Henry was now 30 and saw little prospect elsewhere,
but "had sufficient rank to open a fine avenue to higher responsibilities" in the
army. He had long recognized Stephen's quick mind, enthusiasm for learning,
and omnivorous reading habits, and he had closely supervised Stephen and
Nettie's education since their father died ten years before. He believed
Stephen should pursue without delay "a liberal and professional education,"
and if he chose, for example, "medicine or a chair," that he would be glad to
assist these pursuits here or abroad.

Stephen was now in a quandary. On the one hand, to reenlist might jeop-
ardize his aspirations for a college education and a profession. Yet on the
other, when he thought of "the dear, noble cause and how much hope and life
and energy we have spent for it . . . and leaving my brother and my friends to
carry on the fight," he shuddered. Forbes continued in his letter to Flavilla:
"Oh! How I should sigh for the long, wild rides and merry scouts, and for the
dark forests and green plains of the South, and groan to hear the rattle of a
volley, and how disgusted I should get over the selfish tameness of northern
life, until I should throw my Greek out of the window and kick my Latin into
the fire and go and shoot myself for an insignificant wretch! The fact is, reason
and prudence tell me to go home, and love of country and enthusiasm tell me
to remain, and which shall I obey?"

The latter two won the day. He had no choice, Stephen said, it was "my
duty," and in February, Stephen and nearly the whole of Company B signed
on for three more years. As a reenlistee he was entitled to a bonus in pay as
well as a month-long leave, which he soon spent at home.

He came back to the army in St. Louis with a "strong desire to do my whole
duty" and surprised himself by having "as much energy and earnestness" as he
had back in 1861. But this fresh self-appraisal was now tested by a lull of four
and a half months when he faced boredom and impatience with himself.
From St. Louis he proceeded to Memphis in June arriving on the steamer
Olive Branch, where as officer of the guard he oversaw debarkation and then
went uptown where he found his horse, Robert. "I found him looking splen-

didly . . . the gallant old fellow," he wrote in his journal, and then added, "Oh for the field!"

Soon he began a remedy for his tedium: setting out a regular course of his time and study. He spent mornings studying tactics, company business, and other military matters; afternoons he devoted to Spanish, reading, writing, and recreation; company drill figured largely in both time periods, and at night he read French literature for an hour. Long after the war he remarked:

> It was as easy to carry a little book in ones saddle bags as a pack of cards, and to read or even study by the campfire while one smoked was a profitable recreation which I still remember with delight. . . . Anyone who had kept the solitary flame of his separate intellectual life steadily burning through all the blasts and storms of war might reasonably believe that nothing that should happen to him thereafter could possibly extinguish it; and this, as we all know, is more than can be inferred from the completion of an ordinary college course.

The small library he used during his army years consisted of two dozen books on language, literature, history, and military tactics (see appendix 1). Some were from home, some bought, and some found, he wrote, "almost anywhere that a planter's house had been ransacked by our foragers."[8]

In late June he was temporarily assigned to command Company I, whose captain had gone on thirty-day leave: "If the command of the wildest, hardest, craziest, and most unmanageable company in the Regt. doesn't wake me up and keep me from spoiling, it will be because I don't do my duty." A week later he wrote: "I can see a vast improvement already . . . and now I mean to do my very best to awaken that company spirit and ambition which lies at the bottom of all first class discipline."

Confronting that common command problem of learning to lead men without driving them, keeping a proper distance, and dealing with his own strong conscience as well, the young lieutenant forged ahead. "How strange a thing it is that one man should absolutely rule another, that I should set up my will and my belief against the determination of a hundred others and conquer them by mere strength and steadiness of purpose and resolution."

As that hot, droughty summer wore on, he measured the chasm between his strong idealism and the grim reality of army life. The future lay wholly uncertain. He trembled with dark thoughts of his own mortality, had bouts of deep melancholy, and wrestled with the meaning of fate. "We spend our whole lives in gathering together a little knowledge, when at any moment a singing bullet may teach us more in a moment than all earth knows," and then reminded his sister Nettie that they should both take care "only to fear God and do our duty and then rest."

Forbes provides a picture of the young officer's quarters in Memphis in the summer of 1864:

> I sit in my wall tent . . . in front of my private writing desk.
>
> Over my bed . . . I crossed my guidons forming quite a brilliant background of stars and stripes to my little home. Next on the east, beside the head of my bed, sits a box of ammunition supporting a box of papers against which lean two carbines and my own saber, and then, in the N.E., corner, is a rudely built table covered with the Company books . . . while in front of it are two boxes of ammunition set up for a seat.

Relieved of command of Company I in late July, Forbes was assigned quartermaster and commissary duties, putting to use his good organizational skills. His regiment that summer shuttled back and forth from Memphis to La Grange keeping an eagle eye on General Forrest's cavalry and preventing him from cutting lines of communication to Atlanta. General William Sherman had tersely ordered his commanders "to pursue and kill Forrest." But then Forrest did it again, raiding Memphis in the foggy morning of August 21 with 2,000 picked men, although he failed to capture Union Generals Hurlburt and C. C. Washburn as he had planned.[9]

Forbes's Seventh Illinois was now in Brigadier General Edward Hatch's Division, Second Brigade, Colonel Datus Coon commanding. Hatch's division of 3,000 men was only one of many Union units that were now turned against Forrest, who proceeded to raise havoc from Jackson, Tennessee, southeast to Florence, Alabama, on the Tennessee River (see fig. 1).

Forbes reported six cavalry regiments leaving Memphis on a rainy September 30, 1864. They marched east to Bolivar and then in a northerly arc to the Tennessee River valley opposite Clifton on October 5. It lay before them hazy and blue, framed by the "steep slope of an oak-covered bluff" on which they stood, Forbes wrote, and where the opposite side of the river fell away in a series of "hills, gorges, and bluffs . . . with the rich delicate mantle of forest now just beginning to be tinted with its autumn colors."

Crossing the Tennessee to Clifton via federal transports on October 6 and with Forrest's scent in their noses, Forbes's brigade marched toward Lawrenceburg and first learned of Forrest's passage south. In pursuit they went October 8 and 9 down Shoal Creek, "the most enchanting stream I ever saw," wrote Stephen. "Oh, that beautiful valley! The memory of it comes to me now . . . its clear murmuring stream, its high gorgeous hills, and level green vista between."

Learning that Forrest had crossed the Tennessee River and escaped, Forbes's brigade spent the next few weeks foraging for rations and early in November marched to Pulaski, 38 km (24 mi.) east of Shoal Creek. Here a Union force of 22,300 men was under the command of Major General John Schofield and under the overall command of Major General George Thomas, with headquarters in Union-held Nashville.

Meanwhile Forrest had reappeared south of the Tennessee, joining up with Confederate General John Hood at Tuscumbia, Alabama. Forrest's 5,500 cavalry plus Hood's own infantry and artillery gave Hood an effective fighting force of 32,500, half again as large as Schofield's Union army. Hood's intention was to strike Union forces in southern Tennessee and move on to Nashville.

General Hatch's cavalry, including Forbes's brigade, would act as ears and eyes for General Thomas in Nashville, supplying him with accurate intelligence of Hood's intentions and movements. This mission was critical since Thomas expected 20,000 reinforcements for his command in the next few weeks and wanted to delay and gain time for these troops to arrive. He would try to avoid a decisive battle until they did.

Beginning November 6, when the brigade drove advanced Confederate scouts across Shoal Creek, Forbes and his comrades skirmished with the enemy with increasing tempo. Then on November 19 Confederate cavalry advanced on the west side of Shoal Creek. Moving out on their right flank, Stephen's brigade collided with lead elements of Forrest's cavalry under General Abraham Buford and, with the vanguard of Hood's mounted infantry, eventually cutting their way out against a cacophony of "hideous yells," Colonel Coon later wrote, "such as only rebels can make."

Under brutal weather—rain, snow, freezing temperatures, and biting wind—Forbes's brigade fought the advancing Confederate columns through Lawrenceburg, through Campbellville with its bloody hand-to-hand combat ("hot and heavy, nearly surrounded," Forbes wrote), and then through Lynnville, all the while against Forrest's cavalry under Generals Buford and William Jackson.

General Hood's Confederate army now was clearly moving toward Nashville. Command of all Union cavalry was now assumed by 27-year-old Major General James Wilson. He met Forrest in combat for the first time with Forbes's brigade at Duck River and then twice more on the road to Franklin. Simultaneously, General Schofield, who was ordered to withdraw, nightmarched his army to Franklin bypassing two Confederate corps of Hood's infantry lying in camp by their fires less than a half mile from the turnpike. Schofield's escape was largely due to Hood's poor planning and communications, resulting in a breakdown of his command staff.

Located in the curve of the Harpeth River, Franklin was now the scene of some of the bloodiest fighting and the last great Confederate charge of the Civil War. General Hood ordered an assault at 4 P.M. on an Indian summer November 30, 1864.

Early on in the battle General Hatch's 2,800-man cavalry, including Forbes's brigade with their new seven-shot repeating Spencer carbines, attacked Forrest's cavalry on the Union left flank north of the Harpeth and forced it back across the river. For the first time Forrest was no match for a strengthened Union cavalry and a brigade of General Thomas Wood's Union infantry.

Seven furious Confederate infantry assaults were made by 11,000 men in the main force, two miles over open fields against strong earthworks until 9 P.M. that night. Hood lost 6 generals and 53 regimental commanders killed. Total Confederate losses were probably 7,000, including 1,750 killed, 3,800 wounded, and the remainder captured or missing. Total Union losses were 3,717, including 189 killed, 2,082 wounded, and the remainder missing.[10]

Hood now marched his decimated army 30 km (19 mi.) north to Nashville. Even with General Stephen Lee's recently arrived corps, Hood had fewer than 30,000 men. Reaching Nashville on December 2, Hood laid out his lines in a crescent south of the city (fig. 3) including three attached, and two detached redoubts (small forts) on his left, and then he awaited Union General Thomas's move. Unbeknownst to Hood, Thomas now had an army twice the size of his own.

The weather then turned abominable, freezing rain and sleet forming ice sheets on every surface. By December 14 the weather improved, and Thomas ordered battle the next day. His line consisted of Wilson's cavalry corps on the Union right and four infantry corps and another division on his extreme left. The battle would begin with a diversionary attack on the Confederate right, followed by a massive wheeling attack by Union cavalry and infantry turning and enveloping the enemy's left flank and cutting off escape in their rear.

Union troops were up and ready by daylight on December 15, but a thick fog held up the assault until after 8 A.M. Wilson's 12,000 cavalry troopers marched at 10 A.M. guiding on the right of General A. J. Smith's infantry corps. General Hatch and Colonel Coon's brigade, with the Seventh Illinois, proceeded mounted, including Forbes who acted as an aide to General Hatch this day.

"Presenting a magnificent spectacle," said Coon, the 23,000 cavalry and infantry wheeled forward with guidons and regimental and national flags flying over the hilly ground along the stumpy line of Richland Creek. Offering only token resistance to this sea of bluecoats advancing toward them, gray-clad troopers and infantrymen of Rucker's and Coleman's Confederate brigades were swept away (see fig. 3).

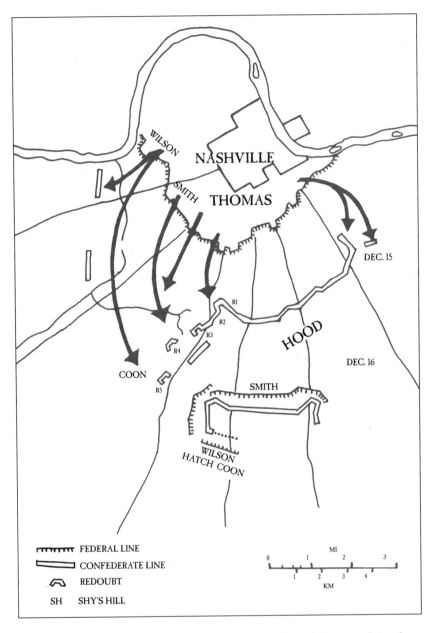

Fig. 3. Battle of Nashville, December 15–16, 1864. On December 16, both Union and Confederate lines were in the Brentwood Hills. (Adapted from two maps in S. F. Horn [1968], The Decisive Battle of Nashville, following pp. 94 and 110.)

As artillery salvos thundered and shells burst overhead, the attacking Union right wing swept across the Harding Pike. Moving on the double-quick, hearts pumping, Forbes's brigade at about noon found themselves at the extreme Confederate left on the flank of redoubt 5. Positioning his now dismounted brigade near a small wood, Colonel Coon shelled the redoubt with his artillery.

General Hatch then ordered a charge with Coon's four regiments including the Seventh Illinois. "Forward!" shouted Coon, and with their rapid-fire Spencer's blazing, the troopers surged into a hail of grapeshot and canister from cannon in the redoubt and musketry from Confederate infantry. In the thunder and smoke "we swept up the hill so quickly," Forbes wrote, "that the enemy had no time to retreat before we were among them," capturing four cannons, 65 prisoners, and battle flags.

Immediately cannon from redoubt 4 on a higher hill began to shell the Union troopers. Even before orders were given the men started for a second charge "like a pack of wolves," and with the buglers sounding the charge, ascended the steep hill to the redoubt. "The first men to reach the ditch before the breastworks jumped in," Forbes wrote, "and waited until there were enough to make a rush over the parapet." More men arrived and hand-to-hand fighting ensued.

Meanwhile, Forbes had ridden forward and heard a wild "Hurrah!" after seeing "a column of gray-clad soldiers streaming out of the fort to the rear," many of whom were among 150 captured with four more cannons and Confederate colors. Coon's brigade then joined infantry units to capture another hill before darkness fell and then dropped into a deep sleep in a bivouac on the Hillsboro Pike. It had been, Forbes would say in his elderly years, "a day I would never forget."

While Forbes's brigade shelled and captured redoubts 4 and 5, Generals Wood and A. J. Smith's corps had swung forward and assaulted the remaining Confederate left wing with infantry and artillery (fig. 4). By evening the complete enemy left wing had collapsed forcing their center and right back 3–9 km (2–6 mi.) to the Brentwood Hills.[11]

That night General Thomas telegraphed his superiors in Washington informing them of the day's victory. The next morning President Lincoln responded, "Please accept for yourself, officers, and men the nation's thanks for your good work of yesterday. You made a magnificent beginning. A grand consummation is within your easy reach. Do not let it slip."

Early on December 16 the four Union corps moved forward and took up new positions in the Brentwood Hills (see fig. 3). Then Union artillery began a heavy and accurate barrage that would continue all day. After another feint

Fig. 4. Battle of Nashville, early December 15, 1864. Looking southwest and showing Smith's Corps attacking Confederate redoubt 2. In the distance other brigades of Smith's Corps and Hatch's cavalry assault detached redoubts 4 and 5 on Confederate General Hood's extreme left flank. Courtesy Frank and Marie T. Wood Print Collection.

on the enemy's right, General Wilson ordered Forbes's brigade to drive the enemy from the hills and attack hard on his rear. During one of these assaults, Major John Graham of the Seventh Illinois was badly wounded, and Major Henry Forbes replaced him. Then Captain Will McCausland from Pecatonica, Illinois, and a friend of both Forbes's brothers, fell mortally wounded, and Stephen wrote: "To fall fighting . . . in sight of victory—God give me such a death when I die!" They captured several hills that day, taking 225 prisoners and many weapons and enemy colors.

Caught now in a continuous crossfire, assaulted all along the battle line in front and flank by infantry, and seeing waves of dismounted Union cavalry charging in their rear, the Confederate line suddenly collapsed from left wing to center to right, and many of the routed Confederates fled over the hills in their rear toward the Franklin pike.

It was a crushing defeat for Hood in which one of the two chief armies of the Confederacy was eliminated, hastening the end of the war four months later. "More than anyone else," wrote General Wilson, "General Hatch should have credit for the active and aggressive advance of the cavalry against Hood's left in front of Nashville." Union losses amounted to 2,558 wounded and 387 killed; Confederate losses were 1,500 wounded and killed plus 4,500 captured and several thousand deserters. These losses were considered relatively low for a battle of this size and indicated this was a battle of striking execution and maneuver, evidenced by continuing interest in it by students of military strategy and tactics. Forbes, for example, as an avid student of Caesar's campaigns in the Gallic Wars, surely reflected on the similarity of the Nashville battle's outcome with that of Caesar's deputy commander Publius Crassus's smashing victory over the Gauls in Aquitania (southwest France) in 50 B.C. using both infantry and cavalry with comparable tactics.

Hood's army now retreated whence they had come, pursued by Wilson's cavalry 160 km (99 mi.) to the Tennessee River over the next ten days. Forbes's regiment took part in this chase, meeting General Forrest's rear guard several times in hand-to-hand fighting. Hood crossed the Tennessee River on December 28 with a shredded command kept from capture and complete destruction only by Forrest and his guards' firm undaunted courage.[12]

Forbes returned to Nashville before Christmas for quartermaster duties and commenced to rejoin his regiment in early January at Eastport, Mississippi. On the way there in charge of the brigade wagon train, some of Forbes's wagons either broke down, turned over, or sat stuck in deep mud. He complained to Nettie: "Oh, if it wasn't for this war! When I think of what I had hoped for myself at twenty, of what I might have been now, and then look at my present situation, I am almost in despair."

Then by early March 1865 the Seventh Illinois was dismounted, and their horses and carbines given to other cavalry units for a spring campaign against Forrest in Alabama. Henry Forbes was promoted to Lieutenant Colonel commanding the regiment. In early April the end of the war seemed near. Forbes and his comrades learned of both General Lee's surrender to General Grant and Lincoln's assassination in mid-April. All of this took time to sink in, Forbes wrote, as he contemplated the "astonishing future which seems to be suddenly opening around us." In closing to Nettie he proudly reported his promotion to captain, making him at age 20 one of the youngest of that grade in the western army. Then in July newly promoted Brigadier General Coon marched his Brigade to Decatur, Alabama, as the Cavalry Corps, Military Division of Mississippi, was dissolved.

Late in July, Forbes learned that his regiment would probably muster out that fall, moving him to ponder his future. The liberal arts had served him well during the war, but now he wondered about a more practical path such as medicine or civil engineering. So he wrote his mother, setting off a spirited exchange among himself, his mother, and Nettie. "It appears to me," wrote Agnes "that law would be more in keeping with your talents and disposition."

If he wanted a classical education instead, he had a fine start; and by teaching part-time, and "hard study," he should be able to finish in two or three years to an eventual professor's chair. "Your success and prosperity lay very near my heart."

Nettie was upset by his mention of medicine and engineering and thought he would exercise the power of his mind better as a professor or lawyer. In fact, Forbes replied, what he had long wanted was a learned profession and teaching languages and literature with an ultimate aim of a professorship in a university. What troubled him was whether after years of wartime service, he could embark on a typical college course among "inexperienced boys." Could he sacrifice and dedicate his life "for the sake of knowledge?" He thought it likely that he could but had not yet made up his mind which way he would go.

Henry had now definitely decided to leave the army. He planned to buy a plantation and go into the cotton business, perhaps in Alabama, and he inquired whether Stephen might be interested in a partnership, although he fondly hoped his brother would speedily finish his education as first priority. In any event, Henry decided his mother should live in his and Jennie's home whenever established.

On October 23 the Seventh Illinois Cavalry Regiment marched to Nashville and were mustered out in early November 1865. They then railroaded to Springfield, Illinois, and were paid and discharged from the army November 17 at Camp Butler, where the war began for them four years and three months earlier. Forbes had fought in eleven battles and a similar number of skirmishes and was never wounded.

Almost fifty years later Forbes reflected on his wartime service, when he, like many others, went in as a boy and came out a man:

> To do the duty of a soldier, each in his place, whatever that might be, was our simple and sufficient code. . . . We knew that we were in the grip of relentless powers; but we had become, as disciplined and experienced soldiers, so well shaped to their control that we obeyed them not only willingly, but loyally, feeling that anything less would be criminal and disgraceful. . . . We knew that we were bound, but we felt that we were free. . . . This personal freedom in some directions, combined with the kaleidoscopic vicissitudes and intensely interesting issues of a life of adventure, made our army service thoroughly fascinating to many of us. We were always eager for action. . . . War was to us cavalry soldiers, at least, not the murderous trade which the stay-at-home pictured it, but a thrilling and absorbing game—the greatest game in the world, in fact, played for the largest stakes.
>
> The effect of all these things upon our character and ideals was simply tremendous. The youngest and most impressible of us were like lumps of hot iron between jaws of cold steel, and nothing that has happened to us since—

nothing that can ever happen to us—can possibly obliterate some of the impressions then made.

The feeling which seized us and drove us to the field of battle and held us to our work month upon month and year after year was much like what one experiences when he gets religion or when he falls in love. . . . Hardship was easy and danger was faced with courage. . . . We could do things which we did not know that we were capable of. . . . We said little of these things, even to each other. . . . Those of us who survived the Civil War in good health and strength, with morals unstained and minds still alert, have had no final cause to regret what seemed at the time the complete wreckage of our plans for life. To us war was . . . at the worst a kind of purgatory, from whose flames we emerged with much of the dross burned out of our characters, and with a fair chance still left of each of us to win his proper place in the life of the world.[13]

And so with that he now began.

SETTING

HIS

COURSE

CHAPTER 3

Medicine and Teaching, 1866–70

Uring December 1865 the Union Army proceeded with rapid demobilization. President Andrew Johnson worked on the hard task of restoration, while the beaten south lacking both capital and currency lay economically prostrate.

On both sides, joyous cries of war-weary people for peace soon gave way to predictable anticlimactic disillusion, impatience, and sorrow when thousands of young men did not come home. Although southerners by and large accepted the result of battle, many would hold a lasting emotional loyalty to their lost cause. Ostensibly freed, blacks would nevertheless be kept down and from voting well into the next century. Few Americans were not haunted by the pointed question: Where to now?

For Henry and his fellow officers the question led them to search for promising investment opportunities. His perusal of newspapers turned up notices of land for sale or rent in the Nashville, Huntsville, and Memphis areas, and frequent sales of mules, other stock, wagons, and harness. By Christmas, however, Henry reported to Stephen that he was finding it hard "to put together conditions of success on any place in this whole country." The only suitable place he had found near Huntsville—about 600 acres—would probably be gone by early January unless bought at once. Moreover, he emphasized that everywhere, corn, cotton seed, and fodder were scarce or unavailable and would have to be brought from afar. Blacks were frequently unwilling to work for whites, who in turn feared a black insurrection. He would try to have definite word for Stephen by January 1, but for now his "worst expectations" for finding a place held sway.

In fact, Henry's scheme just did not work out, and the new year found him back at home. Stephen was in Chicago, with $1,500 in army savings, as a newly registered student at Rush Medical College. He thought himself too old at 21, too short-tempered as a former commissioned officer to think of beginning as a mere freshman in a regular college and working his way through as his alternate plan had required. Instead he would prepare for medicine.

When Forbes arrived at Rush Medical College in January 1866, medical education and practice in the United States were barely on the verge of needed reform. Most American physicians at the time had quite limited and unscientific medical knowledge. Compared with Europe—especially Germany—America also lagged in attention to medical research. American physicians generally depended on practical symptomatic treatment of their patients based on personal observation and experience after the French clinical school. In addition, during the first half of the nineteenth century, the American doctor's kit typically contained materials for "heroic" therapy—purges, emetics, bloodletting, and blistering—measures that were used to reduce fever and inflammation but were unsavory at least, and dangerous at most, to a sick person's health. Fortunately, the use of heroic therapy declined after 1860. Skeptical of regular physicians, some people used self-care. Many turned to irregular homeopathic or botanical physicians who opposed heroic therapy and often showed more sensitivity to the recuperative powers of nature. Finally, surgical operations would remain hazardous and fearsome until physicians both gained adequate knowledge of the causes of infectious diseases and became more proficient in the use of anesthetics during the last third of the nineteenth century.[1]

In 1866 Rush was one of the better of about seventy, mostly private, medical schools in the United States. Rush depended on student fees for all of its income, but operated on a nonprofit basis from the start. The school was founded by an eminent pioneer surgeon, Daniel Brainard, in 1843. It was Chicago's first medical school, and a new building was erected a year later at the corner of Indiana (now Grand Avenue) and Dearborn Streets.

After the Civil War entrance requirements for American medical schools were uniformly low; a high school diploma was not required, nor in fact was one required either by most regular American colleges. Not until 1880 was a high school diploma (or equivalent by examination) generally required for medical school admission. Forbes's class at Rush numbered 274 during 1866–67 and included students from a wide range of ages and backgrounds. About half were from Illinois, and another 43 percent from other midwestern states. The rest hailed from eight other states, Canada, and Norway. Nine students already held the M.D. degree. They were entitled to a Rush degree as well by passing a satisfactory examination and paying a $25 graduation fee.

Requirements at Rush for the M.D. degree stated that the candidate must be at least 21; that he have pursued the study of medicine for three years (including time registered with a preceptor, or practicing physician), and during that time have attended two fall–winter terms of lectures each of four-months' duration; that he have attended clinical instruction at the college and studied practical anatomy (dissection); that he be examined orally in each course by the professor; and that he present an acceptable thesis in some medical subject. An additional year of continuous study and practice was accepted in lieu of a second term of lectures—a reasonable arrangement since the two terms of lectures were identical in subject matter as rote learning held sway.

Eight courses made up the curriculum: physiology and surgical pathology (Dr. Joseph Freer); principles and practice of medicine and clinical medicine (Dr. James Allen); practical anatomy (Dr. R. M. Lackey); *materia medica* and medical jurisprudence (Dr. Ephraim Ingals); chemistry (Dr. James Blaney, new president of the college and founder of the *Illinois Medical and Surgical Journal*); surgery (Dr. Edwin Powell); diseases of the eye and ear (Dr. Edwin Holmes); and obstetrics and diseases of women and children (Dr. DeLaskie Miller).

Student dissatisfaction with some professorial course offerings, and the good instincts of students and physicians alike, led to several other instructional arrangements that helped improve the quality of American medical education. At Rush these began as informal clinical offerings by faculty and local physicians during spring and summer months. They later became more formal, though they remained optional, and included some lectures and recitations on topics not covered in regular courses. Several clinics were provided weekly during the year at Cook County Hospital, the Marine Hospital, the Chicago Eye and Ear Infirmary, and the college dispensary. Clinical instruction was emphasized at Rush and covered such topics as surgical procedures, bandaging, examination techniques, apparatus, post mortems, obstetrics, and diagnosis and treatment of specific illnesses with patients and their attending physicians present.[2]

Because Forbes arrived at Rush just as students were finishing up their regular fall–winter lecture series, he began his own private study of osteology in preparation for the optional spring–summer work starting in March. He was, he wrote his cousin Frances Snow, "extremely well, and pleasantly situated" in his rooming house, where he caught up on his family correspondence and contacted his medical preceptor.

Not that Forbes's adjustment to civilian life was easy. Far from it. He found that "the very structure and operation of civil society," he later wrote, "were more or less a puzzle and a mystery." Compared with the more tightly organized and controlled army life, civil society appeared to him at once incoherent, ungoverned, and irresponsible. He was utterly surprised, for example, "at see-ing valuable property left on a [Chicago] street without a guard." Moreover, the motives of people in civil life now struck him as "selfish and trivial," and he thought the "business struggle for existence and for personal success posi-tively repulsive." In short, the "confusion of a practical democracy was at first bewildering" and took time to get used to. But in so doing he believed that he and other veterans had a potential advantage: they had learned the importance of their need "to subordinate individual interests to the general good," a mo-tive that would skillfully guide his own journey through a lifetime of public service.

Forbes found spring and summer work at Rush much to his liking. The di-versity of clinical offerings proved first-rate. County hospital wards could be walked, interesting operations could be observed, and he got a running start in practical human anatomy. Also, fewer students in residence during summer meant fewer formal lectures and more hands-on work. This resulted in less pressure on professors and demonstrators, more in-depth review sessions, and an all-around improved learning experience. Students at Rush, he noticed, by and large worked hard, harder than students in Europe said those who knew. Some of his classmates had worked as hospital stewards or surgeon's assistants during the war where they had learned and now eagerly shared improved am-putation techniques—perhaps the most important influence of the war on the medical school curriculum.

Occasionally he got away from Chicago: he spent an enjoyable week in May at Fond du Lac, Wisconsin, for example, attending a trial concerning medical jurisprudence and soaking up some beautiful country as an antidote to five months in the city. On return, he learned to his delight that Henry and Jennie had finally found a home on a farm in South Pass (now Cobden) in southern Illinois. "God grant," he wrote Nettie, "that he may enjoy the happiness in it which he so thoroughly earned."

Regular fall lectures began October 3 but came to an abrupt halt one week later with the sudden death of former Rush president Dr. Daniel Brainard. Dedicated and polished, genial but direct, he was considered bold and cool in surgery in those years before the wider availability of adequate anesthesia. Trained in the east at Fairfield and Jefferson Medical Colleges, and with con-siderable French medical experience and acclaim, he held fame as a medical pioneer, not only in the establishment of Rush and the first Chicago hospital, but also for his lucid medical research papers on the treatment of snake bite and healing of fractures—the first papers from Chicago. At age 54, while preparing a scientific paper on cholera, he died of that very disease, one of al-most 700 people who died that October at the epidemic's peak.

Two hundred alarmed Rush students now protested against inadequate hospital facilities for the dead and dying from cholera. Their protest, however, included voting to suspend classes until December 1 and return home. Forbes wrote Flavilla that he was "disgusted and humiliated . . . by the disgraceful conduct of my fellow students;" he felt that as fledgling physicians they should remain in Chicago. So too thought the *Chicago Tribune:* "They of all men should not fly." The Rush faculty agreed and forcibly said so. The students stayed, even though their protest was well founded.

The preventative value of clean water and careful waste disposal in improving urban sanitation and combating disease had been trumpeted in Chicago since the 1850s, especially by Dr. Nathan S. Davis of the Chicago Medical School, decades before adequate knowledge had been gained of the bacterial role in epidemiology. At his and others' urging the city built its first sewers in 1856; the mortality rate due to disease then dropped significantly in the ensuing six years. Growing apathy, fewer new sewers, an increasing population, and preoccupation with the Civil War resulted in a more intense epidemic and high mortality in Chicago during the nationwide cholera epidemic of 1866. The following year the city council established Chicago's first permanent board of health, with three members who were physicians, and in 1868 the council added to the sewerage system.

Forbes and his classmates heard much more about community medicine from the faculty in 1866–67, and from some of them, more about the increasing need for the application of scientific findings and techniques to medicine. These were the years when a migration began—what in the next few decades would become a flood—of American physicians and students to European (especially German) medical and scientific laboratories and institutes for advanced study. Dr. Joseph W. Freer, professor of physiology and surgical pathology, for example, spoke of his upcoming trip to Europe, where he would pursue studies in cellular pathology. And before attending medical school, James V. Z. Blaney, now president of Rush and professor of chemistry, had assisted physics professor Joseph Henry at Princeton and did independent work in chemistry. He was well-noted for his masterly application of chemical evidence in forensic medical cases, and for his useful chemical demonstrations and laboratory work for increasing numbers of Rush students.[3]

Rush was so successful in attracting students during and after the Civil War, that a new, larger building was required for its overflowing classes from 1865 to 1867. Forbes watched the building go up—the "most commodious Medical College building on the continent," boasted the faculty. It sat north of and adjoined the renovated old college that now held a laboratory, dissecting room, and medical museum. Less easily solved was the perennial need for human cadavers for that dissecting room, a problem that was not satisfactorily resolved in Illinois until the passage of an anatomical law in 1885, finally giving medical schools legal access to bodies of the deceased poor.

Soon Forbes's year at Rush was about up. On a snow-covered, exquisitely moonlit January night, he reported to his sisters being "hearty and happy and busy and surrounded by warm, kind friends" despite having a wretched cold. As usual he found it salubrious after "close confinement to the dry tedious work of [his] studies" to put words freely to the page. He must press ahead now and find other income to further his medical studies and practice, not an uncommon need at the time, even for experienced physicians who often also farmed, speculated in real estate, taught, or kept a store.

Forbes first thought he would rent a place, a small farm perhaps, where he might sell produce—he was vague about it. Henry wrote that fruit farms were rarely rented, and although several in his neighborhood were always for sale, that prices were fairly high and still going up. Yet raising peaches, pears, and apples might prove lucrative, and "you could, after a little, find an abundant outlet for your professional activity." He reminded Stephen that many farmers considered the sunnier slopes of the Shawnee Hill country as among the very best in the United States for fruit growing, so if he could come up with the necessary cash, why not give it a try?

Well, he did, but in a different way: "I take charge of a strawberry plantation at Carbondale," he wrote Flavilla in March, "which will occupy about three months of my time each year and with average good-fortune will . . . convey me through my studies with a thousand or two to start with." He was "filled with the new wine of spring" that April, plowing six of the 700 acres by himself and overseeing the rest. He enjoyed the fine weather and—most tellingly—took in the captivating presence of the plantation owner's sloe-eyed daughter. He had also taken up another interest, as he explained to Nettie:

> I have taken to botanizing lately, and having learned to analyze, am making excursions after flowers almost every day. My little microscope helps and interests me very much, and this, with my pipe and my medical work and my business and other matters before mentioned, keeps me busy and happy. Sometimes, too, I take one of my evening walks alone, travelling off into the country for miles.

This early interest in botanizing was his "first love" and portal into natural history. It would remain with him for the rest of his life.

Under a setting crescent moon in the western sky, and with the evening air in his nostrils, Forbes had "the spell" one night, and soon found himself at the dank edge of a great swamp while a chorus of frogs chugged in the blackness. The gathering dusk foretold, he thought, "the deeper shadows and profounder mysteries of life towards which we are all drifting."

"I could almost wish to die on such a night."

To see and study wild nature and not be overwhelmed by it, he was willing to carry an extra measure of "light and strength," and with this armor, to brave his fate.[4]

Fruit growers in southern Illinois that summer of 1867 glutted the market with berries and peaches for the first time in memory, and prices sharply dropped. After the harvest Forbes began searching for winter work. He almost gained a spot in a Carbondale drugstore but lost it when the druggist thought he was overqualified. Forbes decided it was time to move on.

January 1868 found him snug in Mt. Carmel, some 150 km (93 mi.) northeast on the Wabash River, where he both boarded and practiced medicine with Dr. Thomas Rigg, former surgeon of his old cavalry regiment. When he first arrived in Mt. Carmel earlier that fall he worked in a drugstore, and now he also taught occasionally. "I have for chum the principal of the high school," he wrote Flavilla, "and for fellow-boarder the best lawyer on the circuit, so that with the minister living across the way, I ought to be kept in the ways of Wisdom, Righteousness, Health, and Prosperity without much trouble." Perhaps, but not everyone was paying the doctor on time or without an argument. Forbes related the case of a man who objected to a bill of greater than three dollars a day from a doctor who had recently saved his life. Three dollars a day was what he himself was paid for carpentry work, said the man, and that, by God, was enough for the doctor.

Packing up his doctor's bag, putting aside his teacher's hat, and strapping on his plant press, Forbes headed southwest to South Pass that spring for a visit with Henry, and a summer of botanizing in the rocky ravines and forests and on the bluffs of the Cobden hills. Union County and seven others, plus southwestern Jackson County, make up more than 25,000 sq. km (9,610 sq. mi.) of southern Illinois. The dissected edges of the Illinois Ozarks mark the southern boundary of the penultimate (Illinois) glacial advance and unloading some 150,000 years ago. This land between the Mississippi, Ohio, and Wabash Rivers lies at the same comfortable latitude as southern Virginia. It boasts mild winters and brief snowfalls, relatively hot summers, and compared with northern Illinois features a growing season that is thirty days longer and has ten inches more annual rainfall. With more than 2,000 species of known plants, including almost two dozen wild orchids and the southern swamp cypress, and animals such as the green tree frog, and copperhead and cottonmouth snakes, the region vividly illustrates a northern extension of the flora and fauna of the southern Mississippi valley. Accordingly, 40 percent of the fauna is not found elsewhere in Illinois.

As an apprentice physician with new botanical interests, Forbes was among a noteworthy group of physician-botanists from early America who thoroughly enjoyed hunting for plants, collecting them, naming new species, and sometimes adding promising new ones to the *materia medica*. Besides providing vigorous outdoor exercise and satisfaction of scientific curiosity, as it did for Forbes, botanizing served as a practical way of stocking remedies and prescriptions, the majority of which in the eighteenth and nineteenth centuries were botanically based.

Collecting in the Cobden hills—among the highest in Illinois—Forbes came up with a troublesome fern (he would eventually collect 25 fern species). He sent specimens for identification to Dr. George Vasey, physician, botanist, and first president of the Illinois Natural History Society. Two years earlier Vasey had moved from northern to southern Illinois after his wife's death. He gave up his medical practice forthwith for full-time botanical work, a pursuit he had loved since boyhood. Fortunately, Forbes's letter and specimens reached Vasey just before he was to leave for the Rocky Mountains with explorer Major John Wesley Powell. Well known for his modest and kindly disposition, Vasey urged Forbes not to apologize for asking botanical questions. He was "too well pleased at finding one of similar tastes to my own," adding, "Now that you have introduced yourself to me I hope for a better acquaintance."[5]

Before leaving Rush Medical College, Forbes had arranged for a physician in southern Illinois, Dr. J. Copp Noyes of Makanda (M.D., Rush 1863), to sponsor his additional two years of apprentice medical practice. With apprenticeships becoming more and more a liberal formality, the nature of the work and terms of agreement between preceptor and apprentice were left to them. Naturally the apprentice was expected to stay in touch, report on his activities, and eventually plan to spend some portion of his time working with his preceptor. That time had now arrived.

In mid-September, halfway between Carbondale and South Pass, Forbes reported to Flavilla from the little rural village of Makanda in the "Purple Hills." He boarded with a rich fruit farmer and taught in an inauspicious country school of several dozen students. Under his faithful oil lamp, and the steady ticking of his clock, he found pure pleasure in solitary nights of study. "In the company of my books and my own imaginations," he wrote, "I am thoroughly *myself*. . . . I find that I like teaching and am pretty well fitted for it"—not to mention the money it brought in to support his medical work.

Immersed in work and study by year's end, Forbes's thoughts suddenly turned to his sister Nettie and their mutually neglected correspondence. Married now for a year, and caring for her first child, she gave up teaching, and so, she wrote, had more leisure and yearned to hear more frequently from Stephen. "I agree to give you a letter a week if you want and will take my thoughts as they come, without asking me to choose and pick," he replied. "As for yours, you know that that is the way I like them best."

On New Year's Day 1869, Forbes rambled in fog and a soft-falling rain, reaching a thickly wooded hillside collared with a narrow valley below and the great misty Makanda bluffs looming beyond. There he crossed an enchanted woodland floor as "a hundred tiny little plants peeped out at me. . . . We are upon the very extremest borderland of the unfading tropical regions . . . and I always feel, when in the woods, as if I was nearer to the heart of the great Brazilian Forest than to you, with your icy hills and frozen streams," he wrote Nettie.

He wished he could show her the immense fallen hollow log he had found—one he had noticed before but had not really seen. A brilliant golden-green mat of moss covered it, whose crimson and orange-brown capsules were supported by slender red pedicules. "Within the cavity, microscopic fungi had dotted the dark brown mould with dust of shining silver, and little baby ferns were unfolding themselves in the warm moist air—exquisite!"

As January wore on, his writing dwelt on deeper matters, on the "hushed and brooding spirit of night," on the soul and nature, on time and eternity, and on God and truth.

> I cannot tell you yet what I believe or think with regard to all these things. I have gone deeper and farther into their study than I have ever cared to write or tell to any living being, but believe me, Nett,—I am in the hands of God. . . . I need not say to you that these subjects are more important to me than anything else in life.
>
> This much I *may* say, I think, that I am becoming delighted with our correspondence. It is only simple truth for me to say that I grow better and purer with so much thinking of you, and armed with your letters I defy Fate, snap my fingers at misfortune, and meet all buffeting circumstances with contemptuous indifference.

Fourteen years later he would write her fondly, "If I were asked what two things in my experience have done most to educate me, I should say that they were Mill's Logic and my letters to you.[6]

As the school year closed Forbes felt disenchanted and uneasy with the insular little school at Makanda. He knew that he needed another teaching position, one more progressive and fully engaging, where he had more control of the curriculum. And he looked forward once more to being a part of the

family circle at South Pass—Henry, Jennie, their son, Robert, and Mother—
while he helped himself to a pleasing summer of botanizing and natural science.

That October Forbes believed he had found his position: principal of
Union Public School in Benton, Franklin County, with an enrollment of two
hundred scholars. This time he had full control of the school and the curriculum, as well as excellent community support. "I have three assistant teachers,"
he wrote Flavilla, "who need a [good] deal of training, but who work faithfully and harmoniously with me."

Then he put on his doctor's hat and proceeded to give a reasoned and
lengthy diagnosis of his sister's inflamed and bilious liver. He warned her that
she probably faced an extended convalescence, but soon predicted that she
would make a fine recovery and enjoy good health for many years. (She would
live another forty-one years, to age 87, older than all of her siblings at their
deaths.)

His school progressed well that winter, so well that he decided to "stay
for the summer, opening a select school" with one of his assistant teachers.
Hence he was unable to botanize until mid-July 1870. At age 26 Forbes wrote
to Flavilla:

> I confess I don't know yet which way my life will turn. I have many thoughts of
> preparing myself for the teaching of the Natural Sciences and the Modern
> Languages, two branches of study for which I have a special taste, and may
> eventually permit the pill-bags to fade out of my future. . . .
>
> There is nothing at all repellant to me in the idea of the Practice of Medi-
> cine and much that is attractive, but I have no such strong natural bias in that
> direction as we commonly suppose necessary to first-rate success, while I do
> love to study and teach.

Following his select summer school, he wrote again: "I seem pushed in the direction of my secret wishes, and am more and more likely to become a student
and teacher for life."[7]

Returning from South Pass to Benton that fall of 1870, Forbes struggled
with a tangled skein of ambivalent thoughts and feelings. Friends from Benton, more and warmer than he had ever had before, greeted him. His future
seemed bright with opportunities, he wrote to Nettie, and he thrilled to the
"beauty of earth and sky in this peerless September. . . . O, the world is beautiful and life is dear!" He had to "defy Providence itself and all the laws of
nature [trying] to drive me from the course of life which I have determined
upon." Besides, there is "such a mental hunger, such feelings of isolation [that]
oppress one in a society like this." The people of Benton were salt of the
earth, but most folk never had vivid thoughts, and "there's not a living being,"

in the church he attended, "to whom I can talk frankly and freely of my books and pursuits. . . . I shall die some day from pure spiritual starvation."

This letter smacks of the soarings and complaints of a principled and idealistic young person entering the adult world, yet Forbes was on to something: the proposition that living in post–Civil War America was, as the writer Lewis Mumford put it, "an uphill job." Especially outside the university, one found little sustenance from the community for people who lived for ideas. Consequently, one required strong character, intellectual endurance, and a leaven of patience to stay the course. Forbes had the first two traits in abundance, but would need to work on the third.

By Thanksgiving he claimed some improvement and poked fun at himself as well. Forbes wrote to his cousin Cornelia:

> Am giving my leisure time chiefly to the study of the Natural Sciences, for which I have developed a despotic and troublesome enthusiasm. It drives me out into the woods for weeks at a time; it pushes me into swamps and chases me up the tallest trees, and has even been known to drop me out again; it wears out my boots and tears my raiment, and treats me generally like a very old-man-of-the-sea, and rewards me for my labors with stacks of what Mother calls "brush," some of it rare and a little of it "new to science."
>
> But in the winter . . . I busy myself in metaphysics and the languages . . . sufficiently indifferent as to whether Fortune fills or repudiates [her contract with me].[8]

He was encouraged in his pursuits by the appearance in fall 1870 of his first published writings in the natural sciences, notes on his botanical collections from southern Illinois. They appeared in *American Entomologist and Botanist,* with the support of Dr. George Vasey, editor. Indeed, in an earlier issue, Vasey had described a new species of *Saxifraga* and named it *S. forbesii* after "the enthusiastic young naturalist who first detected it . . . growing on shaded cliffs near Makanda and Cobden."

Forbes's article covered distribution, habitat notes, seasonal appearance, and blooming times, and a few anatomical points for some of the more noteworthy plant species among the 450 he had encountered. His collecting, primarily in Union and Jackson Counties, extended over a vertical range of 762 meters (2,500 ft.), from the low Mississippi and creek bottoms to the Pine and Cobden hills and Makanda bluffs, each with its characteristic flora. During the coming months he would receive congratulatory letters concerning his article from naturalists in Chicago and Philadelphia and at the University of Michigan and Harvard. He had worked hard, doing his best and trusting the result would come of itself. In short, he had learned the art of living with uncertainty. The young naturalist now seemed to sense that he was clearly setting his course.[9]

CHAPTER 4

Natural History, 1871–84

After his article's publication in the fall of 1870, Forbes completed his second term as principal at Benton. He had squirreled away part of his earnings from the previous summer's select school, ready for use after term when he would return to school to study "some of his favorite branches." He spoke and read French, German, and Spanish; he had a reading knowledge of Latin and Italian; he easily digested metaphysics and logic; and he thrived on the study and practice of botany, zoology, and geology—all the while keeping up his eclectic reading habit, beginning at home with Henry, then at Beloit, and continuing during his four years in the army.

Occasionally he thought of going to Boston, New York, or Chicago for further study. Some family members urged him to attend Harvard—much as they (unsuccessfully) had urged Henry to attend Yale before the war—but his choice lay closer to home, at Illinois State Normal University, near Bloomington.

Founded in 1857, the university also housed the museum of the Illinois Natural History Society (1858–71), whose spirited research, education, and survey work had been put on hold during war time. By 1862 its collections were estimated at 60,000 botanical, zoological, and mineral specimens. Eight years later they had been augmented by mountains of partially inventoried material shipped east by the indefatigable Major Powell during his four years of exploring in the Rockies and Colorado River country. Powell had lectured in geology at Normal in 1866 and had been appointed curator (director) of the museum there in 1867. But he had raised the hackles of university and society colleagues because of his prolonged absence, and George Vasey was subsequently named deputy curator in 1869. Normal University at the time was considered one of the foremost teacher-training schools in the country, and its faculty was first-rate. It included Joseph A. Sewall, a Harvard Medical School graduate who had returned to Harvard for postgraduate studies in science with Asa Gray and Louis Agassiz. He was now professor of natural science and a practicing entomologist, and he would later be called to the presidency of the University of Colorado.

Admitted as a special student and supported by his savings, Forbes signed up in April 1871 for chemistry with Sewall, and for reading and elocution with John W. Cook. At that time the Normal University consisted of a single, massive, brick-and-stone, three-story building, 49 meters (161 ft.) long, with perpendicular wings 30 meters (98 ft.) wide and a majestic central belfry towering 43 meters (141 ft.) skyward. Situated on an elevated plateau just north of Bloomington, the school commanded a grand view of the landscape and the small village of Normal with its 1,200 inhabitants. The commodious school basement housed chemical and zoological laboratories, lecture and dissecting rooms, and the third floor contained the Museum of Natural History.

Because of Powell's initiatives, since 1867 the ostensible societal museum had been supported by state appropriations, some of which partially financed his Rocky Mountain explorations. Inevitably, within a few years Powell's ambitious western project had little to do directly with the founding purposes of the Illinois Natural History Society—original research, natural history, surveys, and public education in natural science *in Illinois*. Unable to maintain the museum, pay the curator properly, or publish their findings, and depleted in number by the war, many members in 1869 thought they should relinquish the museum and disband the society. They bided their time, however, until two years later when state aid was again provided, but with the firm requirement that ownership of collections and other property be transferred to the state for its "use and benefit," and that the curator be appointed by the state board of education for Normal. This was accepted and accomplished June 22, 1871, one week before Forbes finished his term at Normal, as the Illinois Natural History Society expired. Forbes no doubt heard from Sewall about this matter, about the importance of the museum's collections both for the state and for Normal University, and the vigilance now needed to maintain them.[1]

Forbes was once again in need of a teaching position for the fall of 1871. With help from the Normal faculty, Forbes learned of and obtained a principalship at Mount Vernon, population 1,200 and located north of Benton at the opposite end of Rend Lake. Fairly accustomed now to a principal's duties, he smoothly scheduled more time for his private study of science, stimulated by his term at Normal, his wide reading, and his growing interest in educational reform, specifically the role and encouragement of the natural sciences in the schools and colleges, and among the public at large, a subject then current in America and in Britain.

During 1871 this subject had generated lively interest in the Illinois legislature, resulting in the crafting of a bill for formal introduction of the natural sciences into an enlarged common school curriculum. Supporters of the legislation wished to lift the study of science from bookish routine, draw students and nature closer together, and promote new approaches and ideas from teachers. In order to be certified, teachers would require better grounding in the principles and practice of science instruction. Early in 1872 the new educational statute was in place to carry out these reforms.

Meanwhile, efforts to redefine and realign the natural sciences in Germany had increased in the past fifteen years. Recently, the "scientific" zoologists, among them Carl Semper (Würzburg), Carl Claus (Göttingen), and August Weismann (Freiburg), were grappling with Darwin's evolutionary theory and busy pursuing ecologically oriented questions in morphology, for example, how an animal's form was dependent on the conditions of its existence and geographic distribution. This approach was called "biological" but by 1868 Semper, at least, had adopted the term "oekologie," coined two years earlier by zoologist Ernst Haeckel. These men were helping to put a more scholarly shine (Wissenschaft) on zoology, long associated with either medical education or general natural history.

Forbes had recently embarked on his own independent study of natural history. He also came up with some practical suggestions for those who would try the same, and particularly for teachers who could not afford formal retraining. His guiding motive—after Agassiz—was to study nature and not just books. This meant getting familiar with common animals and plants of the local woods, fields, and waters; gathering, preparing, and studying specimens; and learning fundamental botanical and zoological classification. The important point was to gather facts from careful observation, to move from the easy and concrete to the more abstract and difficult, with an aim of framing generalizations through inductive reasoning. Once teachers gained better development of their mental and logical powers, they would be poised to "guide and suggest" student efforts, instead of making intrusive attempts to force and control. Above all, Forbes would continue to urge students to study plants and animals *in the field,* where "a thousand perfect leaves and flowers lie beneath his eye, and the songs of birds and the hum of a million insects fill his ear," as well as making precise comparative studies of representative specimens.

As his spring term in Mount Vernon wound down, Forbes inquired of George Vasey whether there might be suitable work for a young naturalist in the Illinois Natural History Museum at Normal. Vasey, acting curator in Powell's stead, was not able to say, but having just been appointed botanist in charge of the U.S. National Herbarium, he said he would speak with Powell in Washington about Forbes's request. The major had been trying to raise more appropriations from Congress for his current Rocky Mountain fieldwork. Federal money had become available only since 1870 for Powell's work, and he now required support for fiscal year 1872–73, primarily for finishing topographic mapping of the Colorado River. Beyond that, Powell's professed goal was to secure more dependable, continuing funding for urgent scientific work in a little-known region, a task complicated by the fact that Powell's was only one of four independent western surveys currently supported by the federal government.

In mid-April, Vasey wrote Forbes frankly that Powell was undecided about what to do with his connection at the Normal University museum. Nevertheless, Vasey had recommended Forbes to Powell for a naturalist's position at Normal, and said that Forbes "was the proper person for that place." Moreover, he wrote that the major needed a botanical assistant for his upcoming work in Arizona and would provide transportation and subsistence. It "would be a fine chance to see something of the world and to enlarge your knowledge of the Sciences," but Vasey added that Powell had to know right away if he could come. Forbes showed polite interest, but said he had to return to Illinois by October 1, a commitment Powell could not make.

Early in May, Vasey astounded Forbes by asking whether he contemplated making an effort to secure the curatorship at the Normal museum. If so, he advised going to Normal to "bring such influences as you can to bear on the subject" and being prepared to make application with the board of education at their June meeting.

"I had not seriously thought of applying for the curatorship of the Museum," Forbes wrote back, "did not know in fact that Powell intended to resign, but I should certainly be very much pleased to obtain it. It would be quite in the line of my purposes and would profit me much in every way." He would free himself from teaching at Mount Vernon "for any really first rate reason."

Letters crossed, and Vasey's next one arrived saying Powell had a sick assistant and needed some geological help for Arizona. He would pay $50 per month and travel. Forbes wrote in return that he now had difficulty in seeing how he could go to Arizona at all. Not only would he have to cut loose from teaching, but he would likely jeopardize his getting the museum position too since he would be absent until well into the fall, and possibly beyond.

"I am anxious to see you in the Museum of the University, and I think you can obtain the situation," Vasey reassured Forbes.

Early in June, Forbes reaffirmed his strong interest in the curator's position. He believed that an active resident curator at the museum was essential. Students and teachers in Illinois needed good study collections with well-identified specimens to help satisfy both the newly expanded curriculum, now including the natural sciences, and the certification of teachers. Only a resident curator could oversee that crucial task. Forbes wrote that he would arrive in Normal about June 15 and appear before the board of education at their June 26 meeting. When he learned that Vasey had recommended him for the post to the influential Joseph Sewall, his confidence soared. On June 26 the secretary for the board received a letter of resignation from Powell, and two days later Stephen Forbes, age 28, was named curator. His vagabond days appeared to be over.

The first major task for the new curator was clearing up the confusion as to who owned what among the accumulated collections, some of which, according to Vasey and Forbes, were lower in quality or quantity than estimated earlier by Powell. Some were arguably Powell's property, since early expeditions were, in large part, paid for by him. In addition, he had amassed other private collections before coming to Normal. These had not been adequately identified when he added them to the museum in 1867, yet Powell removed bountiful amounts of collections and sent them to Washington when he left in August 1872. Finally, Powell's collections from 1870 on, when his work was supported by the federal government, were claimed by the Smithsonian Institution, despite insistent demands made by a Normal University committee that *all* Powell's specimens be returned to Illinois. The entire episode left a shadow on Powell's reputation at Normal University for years. It was partially lifted by the his goodwill return of some earlier collections to Normal and by a gift of a superb duplicate collection of Ute and Shoshone Indian handicrafts to both Normal and Illinois Wesleyan Universities.

Forbes found more than enough business at the museum to keep him up and running. Most pressing was the sharply increased request for representative collections of animal, plant, mineral, and rock specimens from all quarters of the state. The addition of science to the curriculum and associated reforms, Forbes wrote, had "powerfully stimulated the study of natural history throughout the state." He sent out any available duplicate specimens but his supply was limited, so he called for help from teachers and students themselves. What they needed, too, were correctly identified and named specimens. To accomplish all this he asked that they make a variety of collections during one or, better, two seasons and send them to him at the museum. A circular announced his plan: "I will undertake to name, select, arrange, and redistribute in such a manner as to give each school participating in the work the benefit of a judicious selection from the whole number sent by all. Good specimens in all branches of natural history will be acceptable, and directions for preparing and shipping them will be sent upon application."

It was a shrewd move: learn natural history by doing. But not surprisingly, only the most dedicated came forward, just enough to keep interest up, encourage others, and allow Forbes to work out the inevitable hitches.

Forbes was happy that fall, occupied with his new position, relishing the change of scene, and inspired in a thrilling way. Back in Mount Vernon the year before as principal, he had tried to hire a well-qualified young woman for a teaching position at the high school but failed. Clara Gaston, 23, was born in Ohio, now lived in Normal with her family, and at that time was a senior at Illinois Normal University. Forbes next saw her as an occasional visitor to the museum at Normal. Forbes picks up the narrative and shows his romantic side:

> Then came the fateful evening when we met at one of the houses in the little town. Many people were there but I saw only her, with her blue eyes, her brown hair, her complexion of pale ivory with a delicate tint of rose on her cheek, her white lawn dress with flowing sleeves. . . . The sweet kindness of her look and bearing made it easy to talk to her, and I ought to have seen her to her home, but I did not dare to take the risk. Indeed, I did not need to, so far as I was concerned, for she went with me to my room, a captivating image floating in the air before me, and I realized that I was in love.[2]

Clara graduated from Normal that summer and in the fall began teaching high school in La Porte, Indiana, 40 km (25 mi.) west of South Bend. She and Stephen began an earnest correspondence. By Thanksgiving he knew he must see her during the holidays. He wrote saying he would like to call on her at Christmas. "I shall certainly be glad to see anyone from home," she replied, "and hope you will call." And so he did, spending several days with her, each of them in rapt attention to the other. Later, out of her presence, he marveled at how easy he felt with her, how pleasant and gracious she was, and how her unselfishness beamed through. He learned that she was the oldest child in a large family whose circumstances led her to do nursing work for several years before attending college. "Both of us," Forbes wrote much later, "thus got the habit early of an overpowering solicitude about something else [other] than our own pleasures."

Speaking for himself, though, Forbes did have things he enjoyed. Frequent walking was one, cigars another, running in the evening yet another. And high on his list—writing to Nettie. In March 1873 he wrote her of Clara: "She is a very dear friend. . . . she writes me letters which are a dream of joy. . . . I have been to La Porte to see her once and shall soon go again. . . . She is brave and earnest, and generous and true, sometimes gentle and sometimes gay, sensible and wholesome." Now again to Nettie after a second visit in early May: "*We love each other!*" And finally in late May: "She will be my wife."

Forbes would propose that they marry during the Christmas holidays. He expected to be reappointed at the museum that July, and he would request a two-year appointment. He figured they could live comfortably in Normal on $1,200 a year. As for his future plans, he wrote Nettie, "My leisure time I shall use to put myself up to the front in some single line of knowledge in technical science, and shall then aim to do original work. The idea of Harvard I abandon." After two years or so he felt that he should "be able to command a place as professor of natural history." (Actually it would take nearly a dozen.) Word concerning their engagement resulted in a tea at Professor Sewall's house, after which Stephen effused to Nettie, "If happiness is indeed the reward of goodness, I am the best man in all the world."

Clara and Stephen would marry on Christmas Day in the modest Congregational Church in Normal before a full house. Clara, wearing a simple traveling dress, was attended by a bridesmaid, and Stephen, in his best mufti, had a medical student friend act as best man. Immediately afterward they left on the train for a few days in Chicago; Stephen would return for Illinois Teacher Association meetings in Bloomington before New Year's.

Collections of specimens had continued to arrive at the museum from Illinois teachers and students. But Forbes believed they could do even better. That fall he had laid out his compelling argument for the benefits, methods, and means of teaching natural history in the public schools. His article in the *Illinois Schoolmaster* stressed the expected improvement of students' observation, judgment, and memory, and above all the attainment of discipline through studies in the natural sciences. He urged bringing together scientific men from museums and educators from the schools to guide and upgrade these studies. "Not only would purely scientific investigation and discovery receive a much needed onward impulse, but we might reasonably hope to rescue the schools from that curse of superficial teaching which now threatens to make the popular study of natural history a scoffing and byword." And it would get the students outside.

Forbes then contacted leading teachers in Illinois. They and other academic friends of natural history joined in late 1873 at Bloomington to organize a new society dedicated to the collection, study, and exchange of specimens and to the rational study of nature. After "animated discussion" this "large and earnest" group adopted a constitution and elected university president Dr.

Richard Edwards as president and Forbes as curator of the School and College Association of Natural History.

More than forty schools and colleges would eagerly participate in societal work during its first two years. Teachers and students would be busy afield: they would send some three thousand specimens to Forbes, who then processed and redistributed them equitably among the society members. In addition, he sent out sets of selected marine specimens in alcohol, "illustrating the sub-kingdoms, classes, and leading orders [of organisms], representations of which are otherwise beyond the reach of Illinois schools." Early in 1874 Forbes also provided all participants with an eleven-page booklet giving directions for collecting and preserving natural history material, with the guiding maxim that "whatever is worth doing at all, is worth doing well." The society remained active for almost ten years, contributing as well to summer schools of natural history in Illinois and to the Illinois exhibit at the U.S. Centennial Exposition in 1876.[3]

As curator of the Natural History Museum at Normal University, Forbes now supervised an institution that had played a prominent role in zoological and botanical work in Illinois since the 1850s. Before 1870 there were studies on the fauna by such dedicated naturalists as Robert Kennicott, R. H. Holder, Cyrus Thomas, and Benjamin Walsh, and on the vegetation by George Vasey and Frederick Brendel. Walsh, a classmate of Charles Darwin at Cambridge, and Dr. William LeBaron served as the first and second Illinois state entomologists, respectively, providing insect work through that new state office during 1867–75. These men were all experienced field naturalists and contributing members of the former Illinois Natural History Society, the formation of which was first proposed by Thomas.

Now a professor of natural science at Southern Illinois Normal University, Carbondale, Thomas was a key figure in the early organization of natural science in Illinois. He was a kindly and ambitious man of average height, with a long eager face, somewhat bulging eyes, and a prominent downswept nose giving him the appearance of being on the verge of saying, "Really?" He lacked college training but studied law privately and was admitted to the Tennessee bar in 1851 at age 26. He then practiced law for thirteen years, during which time and looking for a new challenge, he took up entomology and after diligent study, specialized in the Orthoptera, eventually becoming the premier American authority on grasshoppers and a Lutheran minister besides. Not content with just naming, classifying, and describing insects, Thomas early on showed a strong practical bent. He took great pleasure in giving sage advice to farmers about insect pests through agricultural papers and periodicals.

"Natural history," he wrote, "should be studied for the practical use made of the knowledge obtained." As early as 1861 Thomas had pushed vigorously with a detailed plan for a continuing natural history survey of Illinois. Moreover, like Forbes, he eagerly supported the addition of the natural sciences to the common school curriculum, had a strong love of nature, and applauded Forbes's efforts to supply schools and colleges with cabinets of natural history specimens for more effective, rational scientific study.

Forbes would later write that, under the pervasive influence of Charles Darwin, Louis Agassiz, and Thomas Huxley, "a transforming wave of progress was sweeping through college and school, a wave whose strong upward swing was a joy to those fortunate enough to ride on its crest, but which smothered miserably many an unfortunate whose feet were mired in marsh mud. This wave reached central Illinois in the early seventies . . . the period of the return to nature in the study of science."

This wave of progress was just one example of the relentless growth, change, and upheaval in practically every field of American activity that Forbes and his colleagues would experience over the next three decades, while the schools, higher education, and traditional natural history were shaken and transformed beyond their wildest expectation.

In response to the challenge, Forbes arranged for a summer school of natural history to be held at the Normal museum in July and August 1875. Inspired by Agassiz's Anderson School of Natural History on Penikese Island, Massachusetts, two years before, Forbes picked up Agassiz's standard for his students, "to make nature as it surrounds us its own text book." Arranged primarily with general courses for teachers, provision at Normal was also made for advanced or special studies for other qualified students. Attendance was limited to fifty students; seventy applications were submitted from forty-two counties in Illinois. Forbes set tuition at $15 for the four-week course.[4]

Believing that for optimum learning "a general knowledge of the whole is of more value than a special knowledge of a very little," Forbes and his staff carefully selected specimens for microscopic study and dissection that were typical, ones whose characteristics were true "not of the species or genus only, but of the whole class . . . or else furnished notable exceptions to statements about [the] larger group." Freshwater collections of invertebrates and fishes came from Illinois lakes and rivers, marine specimens fresh from the sea between Portland and Buzzards Bay on the New England coast. Students studied all orders of insects as well as numerous birds and mammals. The botanical course featured flowering plants, ferns, mosses, algae, and fungi. Lectures closely followed laboratory work, and fieldwork included excursions in selected habitats.

Instructors for the school were: entomologist W. S. Barnard, botanist and mycologist Thomas J. Burrill of Illinois Industrial University, Forbes as director, Sewall of Normal University, state entomologist Cyrus Thomas, and zoologist Burt G. Wilder of Cornell University. Wilder, who had assisted Agassiz at the Penikese school's two summer sessions, was of the opinion that a greater variety of marine specimens was available for study at Normal. He also delighted Forbes and the class with a surprise shipment of the famous amphioxus from Naples Marine Zoological Laboratory. Forbes wrote of the summer school: "Our natural history school has been a complete success, and has lifted us all more than any year's work. . . . The amount of work done was tremendous, and yet it was so new, so varied, and intrinsically so interesting, that the students found themselves refreshed and rested rather than worn out. . . . This promises handsomely in every way but the financial. . . . It is the first step in a very long and hopeful way; and like other first steps, it will cost."

Indeed, the four Illinois staff members in the course worked for nothing and paid their own expenses. Forbes later wrote that "it should be remembered . . . that this was a time when college men, as a rule, worked like drayhorses and were paid like oxen." Aggravating these realities was the dire economic slump coursing through the country since 1873, which was felt in Illinois throughout the decade, chiefly by overproduction of and low prices for farm products, tight money, and high debt. Forbes himself would suffer a $200 reduction in salary, a poor reward for his new extra role as instructor of zoology at Normal. The arrival of his first child, Bertha, in late 1874 made the loss of income seem especially acute. Thomas Burrill, who would be a long-term colleague at the University of Illinois, had it right when he wrote irately to Forbes that the state expected much work, extra collections, and special effort from the faculty "without a cent of money" offered in return.

To capitalize on the enthusiasm generated by the first summer school, Forbes made speedy plans for a second session in the summer of 1876. His attractive four-page brochure described the objective of the school as an attempt "to encourage and assist a more general, systematic and intelligent study of nature." Special arrangements could be made at this time for students to pursue even geology and chemistry. The maximum number of students for the course was increased to seventy-five.

But it was not to be. The extensive work required for a comprehensive museum exhibit at the Centennial Exposition of 1876 in Philadelphia, Forbes wrote, "derailed plans for immediate continuance of the school." The state board of education requested that Forbes prepare an appropriate exhibit illustrating the collection, preparation, and exchange of natural history specimens for the benefit of Illinois schools. This took time, yet his typically thorough work paid off: "The fact that it was thought worthy of a medal by the board of judges," said a state agent, "is an indication of its merits, and should afford strong encouragement to the [museum] to persevere in its work."[5]

Not long after assuming the curatorship at Normal, Forbes began systematic field collections of freshwater invertebrates and fishes. During the spring and summer of 1875 and 1876 he added numerous species to Illinois faunal lists, particularly crustaceans, about which Illinois naturalists knew very little at the time. He collected in lakes, ponds, rivers, streams, springs, pools, swamps, wells, and even the mouths of drains. Quickly he personally had six new species of amphipods, isopods, copepods, and phyllopods (fairy shrimp). In December 1876 he presented a synopsis with key for some two dozen crustacean species—including ten crawfish—a work that would remain useful in Illinois for many years. The paper was the first in what became the *Bulletin*, a long and acclaimed series of publications on original research in Illinois natural history, which would receive sustained financial support from the state legislature for its preparation and publication. For this auspicious beginning Forbes had the acknowledged assistance of both Professors A. E. Verrill and S. I. Smith of Yale College. Smith advised him on the preparation, dissection, and microscopic analysis of minute crustacean anatomy, an exercise that soon rivaled Forbes's study of gross human anatomy at Rush Medical College.

By 1858 a large majority of the fish species occurring in Illinois had already been scientifically described by pioneer ichthyologists. Some were described from specimens collected in Illinois waters. During the rest of the nineteenth century many of the remaining fish species would be described, a few by Forbes. With seines, nets, traps, hook and line, and from commercial catches, after a dozen years Forbes would collect for his own use and study 87 fish species in 25 families. Needing help for identification, in 1875 Forbes contacted the eminent ichthyologist, David Starr Jordan, M.D., dean and professor of natural history at the College of Science, Butler University, in Indiana.

"I shall be exceedingly glad to exchange fishes with you or to examine and return specimens," Jordan wrote. Forbes's initial study objectives were fish distribution and their foods, but the scope of his work would steadily increase from both practical and pure perspectives.

Little had been done on the foods of freshwater fishes in the United States—and little as well in Europe, according to the literature he had been able to gather—but his would be a rare microscopic study of foods eaten and in what proportions. On the practical side, fish breeders would benefit from knowing more about appropriate foods for native species. And as Spencer

Baird of the Smithsonian Institution and U.S. Fish Commission pointed out to him, the "principal point is to ascertain to what extent they feed upon each other, and how far any may prove to be herbivorous . . . in connection with measures to be taken for artificial culture, [and knowing which] species may properly be reared together." Forbes's work on fish foods, he wrote, is "one of great interest and much economical importance." If Forbes wished, Baird would later be willing to include a published paper on his results in Baird's report to the commission.[6]

In December 1875 Forbes was also appointed to a committee on horticultural entomology by the Illinois State Horticultural Society (ISHS) at their twentieth annual meeting. His keen interest in birds since boyhood made him a natural for looking into the controversy at the time over the role and effect of birds in farms, gardens, and orchards. Most people knew that birds ate many kinds of insects. The trouble was that some of these insects were helpful to man, beneficial predaceous insects that ate destroyer insects of economically important crops. Were birds likely to eat as many (or more) of these beneficial insects as they did injurious ones? Which animals were, in fact, controlling injurious insect populations, birds or beneficial predatory insects or perhaps both? Some believed that certain birds kept down populations of particular insects that would otherwise become injurious, and so these birds should be encouraged to breed or be protected. Still others wanted injurious birds destroyed. Forbes believed "horticulturists should have asked long ago, 'How does Nature fight bugs?'" and sought answers in that direction.

Forbes found woefully inadequate literature on the subject. As with fish, very little was known about bird foods, except for the commonest bird species, "and the assertions made," he said, "often rest upon data which no scientific man would accept as sufficient." Right away he began collecting birds (through judicious shooting) and determining their foods in the best way he knew—by microscopically examining bird stomach and crop contents. This process was time-consuming and difficult because of the fragmentary condition of the contents. It was not unusual for him to spend half a day on a single bird, where his good skills in identifying minute insect or plant parts by microscopy were employed.

As Forbes progressed the full magnitude of the subject loomed large. There was much to know. He needed information on how foods varied with locality, season, year, the age of the bird, and to what extent bird species competed with each other for food. Most important, he needed to know the kinds and proportions of injurious, beneficial, or neutral food species making up the birds' diets. In short, he said, "We must attempt quantitative as well as quali-

tative investigation." Only then could he make an accounting for a fair estimate of each bird species' services and costs. But all of this would take time,
money, and exhaustive work. For now he would make a first approach.

Forbes reported on his work a year later at ISHS meetings in December
1876. His preliminary conclusions were based on an analysis of 220 stomachs
from 59 bird species selected from more than 1,000 individual birds collected
in and around Normal and in northern Illinois from spring through early fall.
Based on the stomach contents and the known habits of the animals eaten,
Forbes estimated that about half of the bird species could be tentatively categorized as mostly beneficial to agriculture and horticulture, about 10 percent
mostly injurious, and close to 40 percent mostly neutral. Robin, catbird, and
indigo bunting, respectively, were good examples for these three categories.
Two-thirds of the bird species, however, scored in all three categories, illustrating the difficulty of obtaining quick and definitive answers to the overall
problem.

Expanding his collections on major compass points in Illinois during all
seasons, Forbes accumulated 1,500 bird stomachs from 80 species during
1877–78. No longer content with merely showing which birds, and how many,
ate what foods, Forbes now explored a new method: a quantitative estimate of
the percentage by bulk volume for each food item, among the entire stomach
contents for each bird. It would take him some time to work this out, hence
his report to ISHS in late 1878 was necessarily brief. He pointed out, however,
that he needed his colleagues' assistance, especially field observations of birds
feeding, perhaps young birds, and better, of birds feeding during insect outbreaks. Data of this kind, and more, would be important to have before any
practical recommendations could be comfortably made for bird management.
Forbes wrote:

> That the balance of nature should be disturbed only after full knowledge and
> searching reflection will be evident to any one who realizes the complicated re-
> lations of living things and the consequent numerous and remote results of any
> changes in these relations. I suppose it is evident by this time, that this work
> cannot be carried forward to success by . . . the few leisure moments of one or
> two very busy men. If it is worth doing, it should be regularly provided for. . . .
> You must also judge whether it is worth doing at all, whether you already know
> enough about the matter, or whether the question is worth settling in the thor-
> ough way I have indicated.
>
> We will undertake it systematically at the State Laboratory if you wish . . .
> as it is work of the kind which the laboratory was instituted to perform.

Following his paper, leading members of the society complimented Forbes
"for accomplishing great work." They appreciated his difficult task, believed

no other man in Illinois was as well-qualified to do the work, and suggested that "it is the best work that has been done in this direction." They finished by unanimously voting "to do all in our power to assist and sustain him." Six months later Forbes was granted $200 per year for this bird project by the state legislature.

The state laboratory Forbes mentioned was created because by the mid-1970s the Normal Museum had become crowded, so crammed with cases full of natural history specimens, that further work on public exhibits and general laboratory work and study was virtually impossible. Immediate changes were necessary.[7] After discussion and correspondence with colleagues, Forbes boldly recommended that the Normal Museum be converted into a state biological laboratory for research and study in natural history. Public natural history exhibits should be transferred to a museum in the new state-house in Springfield, where botanical and zoological specimens would be displayed along with geological collections. Although the new laboratory would continue to supply specimens to the museum and to the schools, its primary functions would be to do original research, to publish scientific knowledge, and to "translate the spirit and method of science with the work of the school." In this way the new Illinois State Laboratory of Natural History would serve scientific specialists, teachers, students, and indirectly, the general public. The public's general prosperity, Forbes believed, was increasingly dependent on "the spread of the knowledge and discipline of science among the people." His recommendations were approved by the board of education, and an appropriate act passed by the Illinois General Assembly on May 25, 1877, made it so.

The former museum now became commodious again after refurnishing and reorganization. Two-thirds of the floor space was freed up when workmen removed excess specimen cases. Tables for microscopy, aquaria, and general work were relocated.

Cases for working reference collections of zoological and botanical specimens, skeletons, skins, pinned insects, and herbarium specimens—all amounting to more than 4,000 Illinois species—were put in place. Forbes requested and received from Normal University new microscopes, breeding cages, and new library books, soon bringing his holdings to 475 bound volumes. Special care was taken for storage of 3,000 marine specimens in alcohol and for collections of fossils, minerals, and western U.S. birds, mammals, and plants. Forbes proudly announced access at all times, for a small fee, to advanced scientific and medical students as well as teachers for special study in such subjects as general botany and zoology, comparative anatomy, histology, and microscopy.

Forbes's and his colleagues' work through the museum at Normal, the

Summer School of Natural History, and the new state laboratory is a fine example of a uniquely enhanced natural history tradition with a public orientation. Changes in the overall character of the old natural history included laboratory experience, critical examination of specimens by dissection and microscopy, active field excursions, study and course offerings for advanced and medical students and teachers, a state-supported biological laboratory with a prominent research function, and publishing of papers—and all were evident in Illinois during the mid to late 1870s. This is the earliest occurrence of such a diverse set of activities for any state, and it illustrates Forbes's direct role and leadership during a major transition in the United States from natural history to a more professional ecologically based science.

At this time in Germany, major zoological research institutes of August Weismann (Freiburg), and Rudolph Leuckart (Leipzig) were somewhat in advance of the United States, producing vigorous "biological" (ecological) experimental research in adaptation, life cycles, color variation, and form of animals continuing through the turn of the century. But descriptive morphological studies proceeded here as well into the early 1890s, with both types of work sustained within the framework of scientific zoology. Yet strongly speculative theorizing about evolution and preaching about it to the general public were widely unacceptable and obstacles to a young zoologist's career. Successful German zoologists, who like Forbes, favored a public connection with their biological research did so with more practical interests like aquaculture and fisheries on the north German Baltic coast. Examples here are Karl Brandt (Kiel), who followed his predecessor, Karl Möbius, and Max Braun (Rostock and Königsberg). These men and others in both Germany and Britain contributed work with expanded governmental support for biological stations and marine studies in the wake of diminishing oceanic fish stocks during the late nineteenth century.

In addition the philosophical doctrine Vitalism, bearing a pre-Darwinian origin, gained British supporters with an optimistic, cooperative view of life. This point of view underlay not only much of the social and educational reform beginning in Britain during the late nineteenth century, but also encouraged interests in ecology, animal behavior, and by the turn of the century a strongly sentimental movement for "Nature Study" designed for young schoolchildren. Organized natural history in Britain also involved strong interests in ornithology and in bird migration and behavior in the last quarter of the nineteenth century.[8]

Forbes had taken an extended trip east in August 1877 to familiarize himself with laboratory and museum operations in Boston, Providence, New Haven, New York, and Philadelphia. He had two children now: son Ernest was born the year before. Clara, without missing a beat, volunteered to handle Forbes's (she called him Alfred) zoology class while he was away that August, and she brought the class successfully through the dissection of the crayfish. On his return Forbes learned of his mother's weakening. Agnes was in his sister Nettie's care in Freeport, Illinois, where he visited her at once. Later in November she continued to fail and died on November 15, age 77. Her thoughts were of her surviving children and her grandchildren to the last.

Curiously, Forbes noticed some rheumatism in his limbs while his mother lingered, and he reflected on her steadfast faith in the hereafter. He wondered to Nettie how he would fare in old age without similar support he could not yet fully accept. "I know what I ought to have," he wrote uncomfortably, "but I shall at least be certain that I have been honest with myself and that I have never sacrificed truth for comfort."

He wrote Nettie again in the spring of 1878, thinking of her and all his blood relatives: "I find my thoughts going back to the war-time and the church time and always leaving me sad and lonely. . . . My Bulletin No. 2 is out, my microscopes are here, my laboratory is finished, my time is filled with my chosen work, the family are well and cheerful—and I have the prospect of a visit from you, haven't I?" She had not visited them in Normal often, but she came down from Freeport that summer, taking time away from her own three children. After her visit Stephen wistfully wrote her again: "It is a longer reach to clasp hands than it used to be," because they had a different dynamic now, plus he had been unexpectedly preoccupied with his work during her visit.

Bulletin no. 2, *The Food of Illinois Fishes* (1878), introduced a series of nine papers appearing over the next five years on animal interactions via their foods, a subject he believed to be woefully neglected. These papers would provide fuel for the research engine of his new laboratory and spearhead his eventual scientific reputation as the leading and most sophisticated scientist during the rise of American ecology.

After a lapse of three years Forbes held a second summer school of natural history during July 1878, that was, he wrote, "equally successful in all respects." The enthusiasm generated at this school led students to form an "Agassiz Society" and prompted them to arouse interest in the society among naturalists around the state. The response was favorable. Subsequent meetings in December and January resulted in the formation of a more general organization: the State Natural History Society of Illinois. Article 1 of the constitution read, "Its field of observation and research shall comprise Geology, Archaeology, and Biology in all its departments." Fifty-two original members were noted, with Amos H. Worthen, curator and taxidermist at the Illinois State Museum, Springfield, as president, and Forbes as secretary. By year's end sixty-six members were listed, ten of whom were professional scientists. The society met in June for field trips, and in February for papers and discussion. Interest held for 4–5 years but then declined when too few professional men remained to sustain formal oral papers, perhaps discouraged by the fact that the society did not publish proceedings. A proposal to diversify the membership by admitting physical scientists was not approved. The society disbanded in 1885.

Construction for the new laboratory finished up too late in 1878 for Forbes to request state appropriations in support of his work for 1878–80; consequently, he had to proceed with a request only for 1879–80 and then enter the usual two-year cycle. "The laboratory bill passed the Senate by a vote of *thirty-four ayes to one no!*" Forbes excitedly wrote Clara from Springfield in April 1879. "How is that for backing? Everyone talks kindly to me in the House, and I *think* the whole appropriation is secure." Forbes had impressed the House with his poised, clear, and thoughtful request for funding, with careful attention to the benefits that would accrue to state science, and the enlightenment and welfare of its citizens. His laboratory probably received about $4,000, including $350 for work on the foods of birds and fish.

What with Forbes's trips to Springfield—to the legislature and state museum—to schools and various research trips around the state, he was away more and more during these early years of their marriage. He wrote Clara frequently, helping to keep away the melancholy that crept in without her presence. She set his tone, his thoughts, and "his being." "It is my best and purest ambition," he wrote, "to be worthy of you." But occasionally a past love returned, once, for example, when Forbes visited crumbling old Fort Prentiss at Cairo, Illinois, across the Mississippi from Bird's Point, Missouri, where he had been posted as a young cavalry trooper in 1861–62: "I cannot write love-letters from these old fields. They still smell of powder smoke. Where the old flag has waved so gloriously, the flutter of a handkerchief is not so great a matter. . . . And yet I do love you. Until I came [back] here, I had forgotten that I ever loved anything else. All this means is that you missed the first great passion of my heart."

They had two more children soon after, Winifred in 1879 and Ethel in 1881, bringing their tribe of "little folks" to four. Good things sometimes come in twos, Forbes noted to Nettie on Winifred's birth—his brother Henry, putting fruit farming behind him for good, had found a principal's position "at last," in Delevan, Illinois. Henry found even a better spot in 1880, at Polo, only 45 km (28 mi.) south of the old Forbes family farm in Ridott on the Pecatonica.[9]

Meanwhile, Forbes's recent funding allowed him to finish and publish five analytical papers during 1880 in the *Bulletin of the Illinois State Laboratory of Natural*

History. The work on birds covered the thrushes (nine species) and the blue-bird, and it was especially anticipated by the United States Entomological Commission, wrote Cyrus Thomas. The former group of birds included the three most important species for agricultural and horticultural interests: the robin, catbird, and brown (wood) thrush. As part of their diet, thrushes and bluebirds ate both injurious and beneficial beetles, requiring Forbes's famili-arity with fragments of beetle mouthparts in bird stomachs. He first dissected and mounted on slides the mouthparts of the Illinois beetles. He then care-fully drew, counted, and recorded "the teeth on their mandibles, the hairs on their tongues, the dimples on their chins," and crafted an analytical key for identifying beetle remains. The work was difficult, tedious, and at times dis-couraging, making it, Forbes wrote, "one of those questions which . . . must be studied chiefly in the public interest." That interest related to the direct economic injury by birds to grain and edible fruit, and by insects to crops, the latter estimated at tens of millions of dollars each year. The important task still was to determine which birds were materially beneficial in faithfully checking injurious insects or insects that could be potentially injurious if their populations irrupted. This work, he pointed out, also had "considerable sci-entific interest. . . . Since the struggle for existence is chiefly a struggle for sub-sistence, a careful comparative account of the food of various competing species and genera, at different places and seasons and at all ages of the indi-vidual, such as had not heretofore been made for any class of animals, cannot fail to throw much light upon the details, causes and effects of this struggle."

Robins (114 specimens) ate a diet almost entirely of insects—largely injuri-ous—from February through May, and larger amounts of fruit later during summer. On balance the bird was valuable, Forbes stated, "yet he is not so precious that we need hesitate to protect our fruits from outrageous injury . . . removing that excess of robins which . . . would [normally] fall victim to hawks and owls."

The catbird (70 specimens) arrived later in spring, and left earlier than the robin, precluding serious competition between them for abundant similar foods (a situation that would much later be called diffuse competition). The bird ate fairly equal amounts of injurious and beneficial insects, but the harm it caused to economically important fruits, Forbes stated, was "without com-pensation unless we shall find it in the food of the young." Little credit was given the catbird, especially by fruit-farmers.

The more shy brown thrush (64 specimens), frequenter of shrubs and thickets, ate fewer fruits, and about 20 percent of its food as corn kernels, grains, and seeds, mostly picked out of animal droppings. Injurious insects in their diet balanced beneficial ones, although this bird destroyed much larger amounts of injurious June beetles than did robins and catbirds. Overall, the brown thrush provided a valuable service.

To determine if the dietary differences among these three species were indeed significant, Forbes compared his food preference data with a smaller sample (by 57 percent) of the same three species from an earlier study. Selecting twelve food types that were most different, he took species in successive pairs and figured the sums of the differences of the percentages of these foods, first for the smaller sample and then the larger. The sum of these differences increased for his larger sample, showing that food preferences among the bird species held up and were even more evident with a larger sample. He concluded that a monthly sample of 10–15 stomachs for each species was sufficient for practical purposes. This was a rare example at the time of a naturalist examining the adequacy of his sampling with a quantitative method.

Finally, he reported on the bluebird (108), that "gentle little bird, so universally admired" for its plumage and song, and the "general impression of its usefulness" to humans. Bluebirds ate few fruits, and at least twice as many injurious insects (grasshoppers and cut-worms) as beneficial insects and spiders. Nevertheless, according to some entomologists, the beneficial spiders, beetles, true bugs, and parasitic insects eaten by the bluebird could potentially have destroyed many more injurious grasshoppers and cut-worms, and so the bird was not blameless. Forbes could not confidently accept that easy estimate. Instead, he felt there was as much or even a better chance that the population size of beneficial carnivorous and parasitic insects was large enough for them to furnish a surplus of their numbers as food for the birds. Put another way, the population increase of beneficial insects may have been limited by their injurious insect food supply, not by bird predation. Hence, bluebirds and beneficial insects were competing for the same food supply of injurious insects. Furthermore, the food of young bluebirds was not yet known. The certainties, he believed, were in favor of the bluebird. No one would have adequate reason at present, he wrote, "for throttling the song of the Bluebird in his garden with the hope of increasing thereby his annual store of hay and cabbage." Forbes here was arguing for the potential operation of two forces, predation and competition, in the regulation of an injurious insect population, a subject of keen interest and debate among later ecologists.

Forbes's friend, the Illinois Industrial University botanist and mycologist Thomas J. Burrill, wrote, "I am more than pleased with your triumphant vindication of the bluebird." It was, he felt, "the best application of the doctrine of evolution logically considered that has come to my knowledge."

Forbes emphasized that the most serious destruction to economically important crops was typically by strongly oscillating insect species, species like the marauding locusts, potato-beetle, chinch-bug, and army-worm. Yet *all* animal species oscillated to some degree based on the difference between their birth and death rates. What deserved more careful study, he asserted, were the frequency and amplitude of these oscillations, and their natural corrections. "It is only by working in harmony with them," he wrote, "that we ourselves can help to readjust the disturbed order." Before any widespread attempt was made to control birds or any other animal for a prospective good, for example, it had to be shown that this good "will not be overbalanced by some greater evil. . . . The whole burden of proof is on the side of those who would disturb the fixed order of Nature." More and more Forbes found himself dealing with thorny interactions, those among the activities of wild species and human beings. This was uncharted territory into dynamic ecology and would require a carefully reasoned approach before he made recommendations for "practical use."[10]

And so Forbes prefaced his 1880 papers on bird and fish foods with a theoretical treatment of the interactions of organisms. Food was of central importance—the currency of an organic unit of plants and animals *in action*. Here was where they met, crowded, struggled sharply, and more important, collided with the economic activities of humans. And here was where Forbes focused, putting to use his curiosity and shrewd creativity.

Plants and animals are functionally connected and thus interdependent through their interactions, Forbes asserted. When the numbers, habits, and distribution of organisms are disturbed, these effects can spread through nature until readjustments occur naturally. Complete readjustment is never reached, although it was more "closely approached in primitive nature," where "by general agreement," there was a "settled harmony of interactions" between plants and animals—the traditional balance of nature concept.

Humans, by their presence, numbers, and actions, affect and disturb this "Natural system." Forbes believed that if humans better understood how nature responded to disturbances, how the forces and processes by which such disturbances occur are reduced, aggravated, or even increased, that they might "by intelligent interference" avoid or reduce the negative effects of their actions. They might even enlist nature's help for the remedy and removal of these effects for society's benefit.

Forbes rejected as impractical the idea of learning more about the dynamics of relatively undisturbed wild nature through a long series of observations by induction. Instead he employed a deductive approach with the aid of first principles, that is, he made assumptions about the cause of a phenomenon and offered a unifying theory for what he called "economical biology"—here, the production, sharing, and consumption of food by living organisms.

As first principles he accepted Darwin's arguments that species were the result of the effects of natural selection, and he accepted Herbert Spencer's theory that the species is maintained at a cost to the individual. Thus a species "can not long maintain itself in numbers greater than can find sufficient food year after year." An overly abundant predator species, for example, could severely deplete its prey, leaving less available food for itself, and in turn decreasing its own numbers. Or fewer herbivorous insects could result in excessive plant growth, leaving crowded, stunted plants with poor survival, leading to less food for the insect and even lower population numbers of that insect. Hence, Forbes wrote, "we find that the interests of any destructive plant or animal are identical with the interests of its food supply." He proposed that this common economic interest is promoted by natural selection through mutual adjustments of reproductive rates. The same would apply for a single predatory bird species, for example, preying on multiple beetle species, where the predators' best interest would be in establishing a reproductive rate commensurate with not unduly threatening its diverse food supply.

Without exact adjustment between feeders and food-supplying species, some degree of oscillations would again be expected in population numbers of all species. Those animals and plants least likely to oscillate wildly, Forbes believed, would be those "preyed upon by the greatest number of species, of the most varied habit." Likewise, a predator eating a mixed diet of prey, would have ample protection against its own population suffering possible decrease in numbers if one or several food species were in short supply or even absent. Forbes interpreted a lack of food specialization as an evolutionary adaptation avoiding harmful population shifts. Strongly oscillating populations, on the other hand, exhibited boom and bust populations, potentially putting them at marked disadvantage for sufficient food and illustrating their imperfect adjustment to the environment, including "changes in the organic balance initiated by man." Forbes's verbal model for predator-prey interactions and associated population oscillations anticipates relevant mathematical models and the theoretical dispute about diversity-stability relationships put forth almost a century later.

Forbes depicted the role of human belief and behavior here as men cultivated and harvested their farms, grain-fields, orchards, and gardens. Human purposes and nature did not fully harmonize. People put forward much effort to set nature aside, to substitute their chosen domestic plants and animals for wild ones, and to ensure that their harvest went to them alone. Forbes wrote: "He would cut off all excess not useful to himself, and all that is useful he would stimulate to the utmost, and appropriate to his own benefit." Man had "abundant reason for a belief in [nature's] usual beneficence," but he found nature confining. Wild nature "is not an evil, it is simply insufficient" by itself for all the needs and purposes of civilized man.

Forbes had no doubt that humans would continue to disturb nature, sometimes radically. "But the general tendency towards a just equilibrium will make itself felt, and intelligent effort will mitigate some evils and avoid others." Meanwhile, humans needed "a profound respect" and a "thorough knowledge of the natural order." In short, they needed "conservative action and exhaustive inquiry."[11]

In this spirit Forbes proceeded with his fish studies. His preliminary study of fish foods from the Illinois River in the late 1870s had shown that aquatic insects, crustaceans, vegetation, fish, and mollusks, in that order, were prominent foods for fifty-four species of fish: from minnows to northern pike and largemouth bass, to suckers, catfish, and shovelfish. Virtually nothing had been written on this subject in Illinois. And he uncovered only scattered notes and a few papers on Great Lakes fishes from the rest of the United States. For very young fishes he found no relevant American or European work at all. So whatever he touched showed promise for investigation—for the natural scientist and the practical fish breeder alike. Forbes's work here with the relationship of an animal's morphology and behavior with feeding anticipates much later work by ecologists in resource partitioning.

Forbes wondered with shrewd foresight, for example, whether widely distributed species would prove to be generalist feeders throughout their range. "Do closely allied species living side by side ever compete for food?" What sorts of correlations might he find between preferred foods and the structure of fish mouths, teeth, throats, and gills? He tentatively predicted less differentiation in fish compared with birds for both foods and structures. Certainly his preliminary work showed that a number of fish species, like birds, ate different proportions of a variety of foods. Yet his fond hope was that on the basis of whatever correlations he did uncover between food habits and fish structures, and knowing what foods were present in the habitat, he would be able to predict what a less well-known fish would eat without resorting to a tedious study of stomachs.

Immediately apparent too was the dramatic abundance of tiny crustaceans—called Entomostraca in Forbes's time—in fish stomachs. These were the myriad cladocerans, copepods, and ostracods, species of which lived free in the water, on the bottom, in vegetation, and even in bottom debris. Abundant, too, were tiny midges (Diptera), whose slender reddish larvae were aquatic, the so-called "bloodworms." Forbes used the name *Chironomus* for these animals, but they are now labeled *Tendipes*. How important were these foods among the various fish orders? For different ages of fish within the individual species?

Widening his coverage, Forbes next examined 425 more stomachs from 33 spiny-rayed fish species (Order Perciformes) collected in most parts of Illinois during 1877–80. Three examples will illustrate the intricate diets that Forbes discovered for different sizes and ages of fish in 31 of these species.

First, were the agile little darters, 13 species of which reside in rocky or sandy streams, but more abundantly in swift waters, doing what their name suggests. Seventy adults of these mostly 3–4 inch long species ate almost half of their diet in *Chironomus* midge larvae, and another quarter in mayfly larvae. Two other darter species, living more frequently in placid water among vegetation, enjoyed midge larvae too, and almost equal amounts of minute Entomostracan crustaceans found more abundantly there.

Second, were the perches, represented here by the common yellow perch of rivers, lakes, and ponds, a species depicting a changing diet as the fish matured. Young yellow perch less than one-inch long fed wholly on Entomostraca, and before doubling in length, also added midge larvae and young surface swimming waterbugs (*Corixa*). Entomostraca then decreased in importance; mayfly larvae and larger amphipod and isopod crustaceans served as important side dishes. Finally, adult perch dined mostly on fish, mollusks, and crawfish. Forbes also examined young and adult specimens of the coveted walleyed pike (also a perch), a species easily reaching 20 pounds. Young stomachs yielded Entomostraca and small fish; adults had eaten a number of gizzard shad (*Dorosoma*), a herring-like fish that is also consumed by several other predaceous fish, including the largemouth bass. Fishermen, however, considered shad as a nuisance, and "useless," and left them on the bank to rot. Shad too, ate the ever-present Entomostraca in abundance when young.

Third, Forbes painstakingly analyzed 237 stomachs from 14 species in the sunfish family (Centrarchidae): the large- and smallmouth bass, the rock bass, the familiar sunnies, and the favorite crappies. Their diets are portrayed in table 1, which shows about 90 percent of the total foods that Forbes identified from these stomachs. The foods are shown grouped in decreasing abundance for each life stage. The number of stomachs examined are on the left; the number of fish species appears at the top of the figure.

Entomostraca and midge larvae (*Chironomus*) again stand out. They contributed 92 percent of the food for the youngest fish of all 14 species. (Lesser amounts of Entomostraca and greater amounts of midge larvae would be expected for smallmouth bass from swift waters.) Larger young (1–4 in.) of all species continued this diet, but with generally decreasing abundance of both food items, and the addition of mayflies and waterbugs as food for more active young fish. Notice, too, that both large- and smallmouth bass commence preying on fish at this size.

Table I.
Foods of the Sunfish Family (Centrarchidae)

	Numbers of Species			
Life Stage	Black Bass (2)	Rock Bass(1)	Sunfish (9)	Crappie (2)
Adults (101)	Fish Crawfish	Mayflies Crawfish Fish	Fish Mayflies Mollusks Water bugs	Mayflies Fish Dragonflies Entomostraca Water bugs
Young 1–4 inches long (106)	Water bugs Fish Mayflies Entomostraca *Chironomus*	Water bugs *Chironomus* Entomostraca	Entomostraca Mayflies *Chironomus* Water bugs	Water bugs *Chironomus* Mayflies Entomostraca
Youngest ‹1 inch long (30)	Entomostraca	Entomostraca Mayfly larvae *Chironomus*	Entomostraca *Chironomus*	Entomostraca *Chironomus*

Source: Adapted from Forbes, *Bull. Ill. St. Lab. Nat. Hist.* 1, no. 3 (1880): 62–64.

With the exception of traces, Entomostraca appeared abundantly only in the adult food of the crappie, *Pomoxys*, a fish with "the best *straining* apparatus," Forbes wrote, "the largest, finest, and most numerous gill rakers" of all the adult fish he examined. Mollusks appeared abundantly only in the brightly colored pumpkin seed sunfish, *Eupomotis*, making up almost half of its captured food. Stout, blunt pharyngeal (throat) teeth in this species assisted a molluscan diet begun in young fish 1–4 inches long (not shown). Fish, mayflies, and crawfish are shown serving as important foods among the adult fish. Forbes noted that the percentage of fish in adult diets increased with a corresponding increase in the relative size of the fish species' mouths. Hence, the largemouth bass, *Micropterus*, and the wide-mouthed sunfish, *Chenobryttus*, had eaten fishes making up 86 and 46 percent, respectively, of their diets; other smaller-mouthed species ate correspondingly less.

And finally, Forbes examined the youngest individuals of species from other fish orders, and compared their diets with the youngest spiny-rayed fishes just described. He examined fish from 14 genera in 4 orders, most ranging in length from .38 inch to 1.5 inches. These included 83 juvenile individuals of bow fin, gar, pickerel, gizzard shad, shiner, suckers, and catfish—14 species in all.

Two-thirds of these tiny fishes' diet was again composed of Entomostraca and midge larvae. Diets of several species of bottom-feeding suckers, however, consisted of almost two-thirds protozoa, algae, and diatoms. They also ate Entomostraca, but no midge larvae. Three specimens of gar and pickerel revealed almost half their diet as fish, displaying their ferocious predatory behavior at a tender young age; the remainder of their diet was primarily Entomostraca.

Forbes's manifest conclusion from these studies was "the supreme importance of Entomostraca and the minute aquatic larvae of Diptera as food for nearly or quite all of our freshwater fishes . . . of whose very existence the majority of the people are scarcely aware."

This immediately raised the question concerning the extent and intensity of competition for similar foods that might exist not only among the many young fishes, but even between young and certain adult fishes like the shovel-fish *(Polydon)* with its remarkable strainer-like gill filaments. Forbes raised a hypothetical case wherein gizzard shad were intentionally increased as food supply for their usual predators, the black bass and walleyed pike. He foresaw here the possibility of a significant reduction in Entomostraca especially by young shad (who were toothed unlike adults) and a corresponding effect on valuable young bass and pike—a curious case of the prey causing "through its own abundance, a serious diminution of the very species which prey upon it." Here again Forbes's thought experiment employed the operation of both competition and predation regulating population size.

In effect, attention had to be paid to *all* the conditions of life, including food, that supported adequate population levels of economically important fish. For example, Forbes noted that "free access of fishes to the ponds, lakes, and marshes naturally connected with a stream is a matter of the highest importance." These habitats supported larger amounts of Entomostraca than the main channels of swifter streams and rivers, and lack of access to these food supplies could affect the breeding of fish. Dams and lack of fishways on rivers would materially deny access as well. In typical Forbes style, he offered practical management suggestions for freshwater fisheries, along with important new fundamental knowledge in the biology of fish.[12]

Forbes's work soon became known in Illinois, and eventually he was widely seen as the leading man in aquatic biology in the United States and as the founder of scientific economic ornithology. Yet as often occurs in life, not everyone was happy with the niche he had carved out for himself. He would occasionally hear of grumbling over his tactics, and beyond that, his strategy. Dependent now on state appropriations for the continuity of his work, he had carefully cultivated key legislators holding the purse strings. This grayed his hair, Clara said, and took him away from his favorite work more frequently than he liked, but he knew it was necessary and like everything he touched, he put his best into it. Fortunately, by the early 1880s he had several competent assistants to whom he graciously attributed "so much of the gratifying progress and conspicuous success of the work of the Laboratory." They included: F. M. Webster, W. H. Garman, A. B. Seymour, Cyrus Butler, A. K. McCormick, and Rachel Fell, the last of whom took over the job of making thousands of microscope slides of many items, including Forbes wrote, "the many hundreds made in the study of the food of insects and of young fishes." Appropriations for the State Laboratory of Natural History for 1880–82 amounted to about $9,000.

Kind words from good friends at key times helped, too. Thomas Burrill wrote again in February 1881: "If there has been any suspicion on the part of anyone that you were not doing the most which possibly could be done with the means at your command for the practical natural history of the state and doing it in the most generous and self forgetful way I have yet to find the man to entertain it."

As the 1880–81 fiscal year wound down, Forbes made field trips for fish in the Illinois River, birds and insects in orchards west of Normal, and in June, for the same in an extensive swing through southern Illinois. Afterward he decided to employ dredges, trawls, and tow nets in Lake Michigan, and (by his assistants) in several lakes in northeastern Illinois and southern Wisconsin during the latter part of 1881 for collections of invertebrate animals and plants. This required some instruction and experience, so he wrote to Spencer Baird asking if he could possibly accompany the U.S. Commission of Fish and Fisheries steamer *Fish Hawk* in her operations out of Woods Hole, Massachusetts. Yes, indeed, said Baird. He could join Professors A. E. Verrill and S. I. Smith of Yale; Dr. Tarleton H. Bean, U.S. National Museum; and the prominent botanist W. G. Farlow during August for some Atlantic deep-sea work on the Gulf Stream slope. Although he did not enjoy the sea swells, the work went well, and on return to shore, Forbes went to Boston, where he loaded up with "the necessary outfit of dredges, trawl, towing-nets, sounding lines etc.," for his work in Illinois.

Work on Lake Michigan commenced in late September, covering the shallow inshore water from Hyde Park to Evanston, and then out to 16 km (10 mi.) from shore. Chugging back in a rented Mackinaw boat to Chicago after dark one night, Forbes stayed vigilant, having heard stories of assault and robbery on the docks. "I had my hand on an open knife in my pocket," Forbes reported later, "but didn't see a soul." Needing samples from deeper water, he then trekked to Grand Traverse Bay in northern Michigan in October, and with assistance of a four-man crew and a steam tug, dredged, netted, and trawled in depths of 30 to 102 fathoms. He found the work exhilarating, getting familiar again with the "roving life." "It recalls, with its uncertainty and bustle," he wrote Clara, "its innumerable makeshifts and its little hardships, the old army life." By late October he had finished his work and was ready to come home. Wanting some fresh air and exercise, he left his hotel on Halloween and walked briskly in the woods.

"Found some partridge berries in the leaves and this set me right again," he wrote Clara. "Had a genuine happy botanist's ramble. There is nothing like it in zoology. If I were to study for amusement, I think that I should go back to my first love."

Back home, late in 1881, Forbes faced some roller coaster years just ahead. Still immersed in intensive writing, he grew reflective and philosophical, defending himself against the grind of it, and remaining firmly optimistic as usual. He had four children under 10, the fifth anniversary of his laboratory loomed, and he would soon be forty. Perhaps imperceptibly to him—but not to Nettie—he had definitely changed. She thought he sounded lost from a religious standpoint.

No, Forbes wrote, not lost, "I've just changed." He now had considerable practical work and aims, and assured her that he would "never lose the dear old-fashioned memories of the home where my strength was nourished and the essentials of my character were measurably fixed. . . . Let us encourage one another." She surely needed *his* support for a nagging illness that fall.[13]

Early the following April, Forbes learned that Cyrus Thomas would be submitting his resignation as state entomologist. Thomas had in hand a new position in the bureau of ethnology at the Smithsonian Institution. Tom Burrill wrote Forbes that he was interested, would accept if asked, but said, "I shall not seek the appointment." Forbes felt the same. They also agreed that the State Laboratory of Natural History would be better located in Champaign at the Illinois Industrial University. The School of Natural History in the College of Natural Science was there. Burrill, in fact, held the professorship in botany at the school, was vice-president of the university, and would soon be dean of the college, and D. C. Taft held the post in geology and zoology. The laboratory's future would be more adequately assured in a young growing institution with a stated mission of promoting scientific research, unlike the more restricted teacher-training program followed at Normal.

Burrill actively supported his friend Forbes for state entomologist, as did many horticulturists and farmers, and so, it turned out, did Cyrus Thomas, who in April recommended Forbes to Governor Shelby M. Cullom. The governor formally appointed Forbes to the post on July 3, 1882.

Meanwhile in Champaign, Regent Selim H. Peabody in the course of reforming the College of Natural Science since 1880, had found Professor Taft professionally inadequate, opening up in June the professorship in geology and zoology at the Industrial University. Peabody's first choice—no doubt supported by Burrill—was Forbes, who after some deliberation, declined.

"My work grows . . . in a slow and healthy way," Forbes wrote his sister Flavilla. "It is a great care to me now, for a reorganization is imperative since the State Entomologist's work has been laid on me, and I cannot clearly see the outcome." He was concerned that his situation might be "built upon the sand," and that "a flood is always imminent," since he was not convinced that the state was up to developing science "at this time." He turned down the position at Champaign with regret, he wrote, "because the work expected [including geology] was not all along the lines I wanted."

Peabody's next offer went to David Starr Jordan, M.D. for a chair, but only in zoology. After a visit to Champaign, Jordan also declined. Although he believed the zoological work had fine promise, he wrote Forbes, the $2,000 salary was only slightly higher than his current salary at Indiana and involved more work. Besides, he admitted he had "a perhaps unreasonable prejudice against an unbroken prairie" environment. Casting his net again, Peabody finally snared Benjamin C. Jillson, a graduate of Yale's Sheffield Scientific School, as professor of zoology and geology.

Forbes's prescription for all this administrative intrigue was simple—fieldwork in all parts of Illinois concerning birds, fish, and insects during July through November, the best part of being a field biologist when he could reconnect with nature and recharge his batteries.[14]

Several times in the course of his bird and insect work, Forbes emphasized that the principal injurious insects showed strongly oscillating population numbers. He urged that more be learned about these insects, that other insect species be examined, and that the frequency and amplitude of their oscillations be recorded and charted. He looked for natural corrections of insect outbreaks, for instance, the influence birds might bring in checking extraordinary irruptions of insect populations. Also, he asserted, "We must learn to what extent birds depart from their usual [feeding] practices when confronted by an uprising of some insect species."

To do this Forbes chose J. W. Robison's 45-acre apple orchard, located west of Normal, for a field study. The orchard had been infested for several years with cankerworms, the caterpillar of the moth *Paleacrita vernata,* and Robison, who shared an interest in science, was all for Forbes having a go at it. The orchard looked devastated. Some trees were fully stripped of leaves, others were dead, and most trees were injured. Over a two-year period Forbes observed forty species of birds in extraordinary numbers and choral performance, attracted he would show, by cankerworms. He sampled the birds in May of 1881 and 1882, and compared their foods with those of selected similar species also collected during May from other Illinois locations, but ones that had more ordinary population numbers of insects. The orchard birds devoured moths from mid-April through June, including wingless females and eggs on the ground, and caterpillars on the trunks, branches, and leaves of trees. By chance he stumbled on outbreaks of three other insect species in the orchard: two beetles and another moth species. One of these beetles, the vine-chafer (*Anomala binotata*), was particularly sought by birds.

For the 141 bird stomachs of 36 species, from tiny wrens to bluejays that Forbes examined from the infested orchard, close to 40 percent of their food was cankerworms. Concentrated feeding of the birds on the irrupting insects accompanied a general decrease in the numbers of most other foods usually eaten, and not just a complete neglect of some foods. The birds were, as later stated by ecologists, opportunistic feeders. So in helping to check the superabundant insects, the birds were not simultaneously setting up other possible strong oscillations. In fact, the orchard birds ate as many species of plump caterpillars as usual (other than cankerworms), providing strong evidence that caterpillars were an especially favorite food of the birds. The extraordinary numbers and diversity of birds in Robison's orchard acted as "a regulator or governor of insect life," Forbes wrote, as one of the dynamic "checks and balances of Nature," when they gorged themselves on cankerworms and beetles.

"There is in Nature," Forbes reported in public in late 1881, "what the old physician called a *viz medecatrix naturae*—a certain remedial power, which brings everything back, in the end, to its usual normal state. . . . Food is better than medicine, and the insectivorous animals are more important to us than Pyrethrum powder and curculio (fruit insect) traps."

What Forbes wanted to show was that the organic life of the world could be likened to "a vast and intricate machine—whose various parts, its groups of plants and animals, are no more independent than the parts of a Corliss engine." Birds played their part, he said, but only in relation "to the things and actions which the living plants and animals of the earth compose." To gather and look only at bare facts about birds, or any other species separately, was a serious error frequently made. Facts are "useless and even dangerous until they are correctly generalized," and theories are constructed and offered for use to the ultimate benefit of science and the public it served.

"The economic biologist," Forbes said in a public address in late 1882, "must be able to cross the boundaries between the specialties and link one thing with another. . . . *His* specialty in the *vital relations* of the various organic groups . . . the interactions of organisms, in short." Yet those responsible for this work, he said, should ideally have a corps of specialists to address these goals, since no individual could easily be expected to do the work alone. Progress in economic biology, he believed, required a well-coordinated approach, and he paid thanks for the fine work of his assistants, for state funds, and for the lively interest and help of his colleagues and friends in the State Horticultural Society. Here was the quintessential Forbes, bringing scientists out of their narrow specialties into new connections promising further understanding and rational stewardship of nature. For him their pursuit of science in the public service was its own reward.

The spring of 1883 found Forbes finishing up his manuscripts on fish work from Illinois and Wisconsin. Writing to Nettie, he said his arm "had grown rheumatic with technical composition." No longer did it translate "feeling and thought," rather it "runs with a crank and a train of cogged wheels, and grinds out only the record of facts and deductions from them." He was afraid that he might lose all that he had gained from his old "spontaneous correspondence" with her because of "the dull, increasing grind of bulletins and reports." Still he believed for now that the future was secure. He was rapidly making a first-rate name for himself, especially in Champaign, where Burrill and others kept a watchful eye on him and his good work. Burrill was in close contact with Forbes, whom he had been advising lately on bacterial disease of caterpillars.[15]

But Forbes's finest contribution that year was the clearer light he would throw on fish foods and the Entomostraca-fish connection. Most relevant was current public concern for the common whitefish *(Coregonus clupeiformes)* from the Great Lakes and other freshwaters northward to the Arctic Circle. In 1879 fishermen landed 21 million pounds of whitefish from the Great Lakes valued at more than $750,000 and making up one-half of the total fish catch there. Questions had been asked as early as 1872 about whether populations of whitefish could be sustained in the face of ever-increasing catches. The U.S. Fish Commission had soon after failed to find appropriate funds to complete experimental hatchery culture and rearing of whitefish. When some fry were captured later from the Detroit River and examined, their stomachs contained only diatoms. Fish Commissioner Spencer Baird consequently turned to Forbes for his expertise in determining more precisely their early foods.

Adult whitefish were toothless and were known to feed generally on bottom animals—small mollusks, crustaceans, and aquatic insects, as well as on plankton. Forbes knew that very young fish of two other toothless families, the suckers and the common minnows, thrived naturally on protozoans, algae, and Entomostraca. To pin down the natural foods of whitefish fry in the field meant hunting for elusive young fish during cold and stormy months in the vast open lake where they had lived since being spawned earlier in the fall. Instead he chose to work with fry that were artificially hatched. The basic problem was to identify the first foods of the whitefish at 3–4 weeks old, just before or after the yolk sac of the tiny 13–15 mm (.5–.6 in.) fish had been used up. This was the crucial time; lack of proper food then, whether in a hatchery or in the lake, meant starvation.

During December 1880 Baird arranged for periodic shipments of young whitefish fry from the Northville, Michigan, hatchery to Forbes. The most useful fish Forbes received had been reared in spring water flowing through natural ponds before entering the hatchery. These fish were also fed daily with freshly ground-up amphipod crustaceans. Forbes chemically rendered these fish transparent and microscopically examined them. Of the 340 fish he examined, 47 (14%) had food. This included crustacean fragments, gnats, *Chironomus*, plant fragments, and some Entomostracans. And surprise—on the lower jaws of fish whose yolk-sacs had nearly disappeared, stood two pairs of teeth that appeared useful for snaring minute crustaceans. This reminded him of the gizzard shad—bearing teeth when young but toothless as adults. His results were interesting but still inconclusive.

And so he attempted, he said, "to imitate more closely the natural conditions of the young when hatched in the lake." He supplied a group of fourteen young whitefish in an aquarium with small aquatic animals and plants from wetlands in the Normal area. The foods included copepods (*Cyclops, Canthocamptus,* and a large species of *Diaptomus*), several large cladocerans, and many diatoms and filamentous algae. The cladocerans and *Diaptomus* were too large for the little fish to handle, but all fourteen fish captured and ate the smaller *Cyclops* and *Canthocamptus*. Forbes reported seeing even the tiniest fish pounce on a *Cyclops*, shake it, and kill it. His evidence so far pointed to the smallest Entomostraca as first food of choice, but he wanted an even more natural set-up to cinch his case.

In March 1881 he managed to secure use of a large 750-gallon aquarium in an exposition building in Chicago on the shore of Lake Michigan. Fresh lake water at 42 degrees F. was continuously supplied to this tank from pipes opening two miles offshore, and then aerated and circulated throughout the twelve-day experiment. In addition, Forbes supplied more fish foods from tow-net samples taken from the lake by boat. A fine wire screen at the outlet of the aquarium hindered escape. Approximately 1,200 whitefish fry were successfully maintained in this fashion through the experiment. At intervals of 1–4 days samples of young fish were removed, preserved in alcohol, and later dissected. Intestinal contents were then mounted on microscope slides and permanently preserved.

Tow net samples of potential foods present in Lake Michigan yielded profuse diatoms; filamentous algae; rotifers; abundant Entomostraca—the copepods *Cyclops, Diaptomus, Limnocalanus,* and the cladoceran *Daphnia;* and *Chironomus* larvae. Of a total of 106 experimental fish sampled, dissected, and examined, 63 (59%) had food. Foods eaten included small species of *Cyclops* and *Diaptomus,* together making up 85 percent of their food, as well as rotifers, and *Daphnia.* Only 10 fish had eaten small algae, diatoms, and filamentous algae. Clearly the small copepod species of *Cyclops* and *Diaptomus* were the first foods of choice of the whitefish in Lake Michigan. Moreover, these two copepod species proved to be new to science, and along with five others, were formally described by Forbes in 1882. The two Entomostracan whitefish foods of choice he named *Cyclops thomasi* (adult 1 mm [.04 in.] long) and *Diaptomus sicilis* (1.7 mm [.07 in.] long).

Wrapping up the early natural history phase of his work in 1883, Forbes reported on foods of 319 specimens of the smaller fresh water fishes of Illinois. These included the sticklebacks, silverside, killifish, and true minnows, all generally smaller fish serving as important foods for larger fish species, and often used by anglers as bait. Not surprisingly, some two-thirds of the 24 species in the listed groups featured some *Chironomus,* Entomostraca, or both, in their adult diet.

Much of his effort was spent on the true minnows, or Cyprinidae, since, he remarked, "in number and variety of species it is much the most important family of freshwater fishes," and incidentally one of the more difficult families for identifying species or genera, even with adult specimens. Cyprinids (like suckers) lack teeth at all ages, and the very young eat protozoa, algae, and Entomostraca, "almost indiscriminantly," Forbes noted. The adults were more interesting and can be summarized briefly.

Based on the length of their intestines, and the presence or absence of hooks and rugged grinding surfaces on their pharyngeal (throat) teeth, the adult cyprinids could be put in distinct groups, but only the length of intestine was connected with a significant difference in diet. The six species in the first group had relatively long, coiled intestines. They consumed a remarkable 50–75 percent of their food as mud and rich organic matter, and only about 13 percent as animal food. In contrast, the nine species in the second group had much shorter intestines, and consumed essentially no organically rich mud. Animal food made up a much larger 75 percent of their diet. This work on the cyprinids clarified and substantially extended some earlier brief statements by the ichthyologist E. D. Cope regarding foods and the structure of the digestive system of the true minnows.[16]

Meanwhile in Champaign, Benjamin Jillson's chair in zoology and geology was in jeopardy. At the end of his first year some of his students were questioning

Jillson's fitness for the position. The ire of even a few dissatisfied students worried the faculty, since there were only about eighteen students in the School of Natural History at the time, students who were preparing to be practical geologists, collectors and curators, or superintendents of scientific explorations and surveys. Dean Tom Burrill held a steady wheel, but in early 1884 Jillson submitted his resignation effective at the end of the current academic year. Forbes had remained informed of this situation through Burrill and Parker Earle, who was a board of trustee member at the Industrial University, a horticulturist from Cobden, Illinois, and a person of some influence with the state legislature. Burrill wrote early in March that he hoped Forbes would now "cast your lot with us." Forbes agreed, but only for a professorship in zoology and entomology, as in fact he had recently proposed to Earle. Moreover, he would retain his position as state entomologist. The directorship of the state laboratory was not as easily addressed, but Forbes fully intended to keep it and he now proceeded to work out an acceptable plan. Final approval by the Illinois legislature was needed for the transfer of both positions, a result that Forbes quietly and confidently anticipated. The entomologist's position had historically been mobile, he had founded the State Laboratory of Natural History, and the signals from Springfield were positive.

In mid-March, Earle notified Forbes that the trustees had voted unanimously to accept his plan and requested Regent Selim H. Peabody to offer Forbes the professorship as he had described it. The trustees, Earle wrote, "hoped and believed that the Natl. History Laboratory will follow you to the University. If not another can be created." Obviously they did not want to lose Forbes again if they could help it.

In the midst of these negotiations David Starr Jordan wrote Forbes from Indiana University: "I do not know whether you set much value on College degrees, but whether you do or not, we should be glad to give you the degree of Ph.D. at the next Commencement, in recognition of your scientific work. If you do care for the degree, will you kindly send me a copy of some one of your published papers to be deposited as a 'Thesis' in the library?"

Forbes chose to submit his recent work on insect irruptions and birds in Robison's orchard—"The regulative action of birds on insect oscillations"—as a thesis. Granting of the doctoral degree was approved and given to Forbes that June in Bloomington. This was Forbes's first college degree of any kind, coming eighteen years after he first registered as a student at Rush Medical College, and it pleased him immensely. Many years later (1925), Forbes commented on his receipt of the Ph.D. in this fashion: "Of course nothing of this kind would be thought of now, but, as I remember, it was not so very unusual then." Regardless, he had certainly earned it.

However briefly, in late April Forbes gloried again in his blessed writing connection with Nettie, fleeing the present and going "back to that free past, when my soul and my brain were my own," he lamented. He feared that he would not work and feel again "as nimbly and cheerfully" as he did back then when "trying to lift myself by my own ears." Now he would have a "new harness" in Champaign and he wished he felt "as ready for the new work as a man of forty (Think of it!) ought to feel" for a new and important task. He hoped to take his scientific accomplishments at Normal and add to them the zoological and entomological teaching at Champaign, "organizing the whole eventually in a way to make each of my three hands wash the others. I wish the opportunity had [been] offered ten years ago."

Shortly thereafter Peabody offered Forbes his coveted professorship effective September 1, 1884, but said that "more time could be had" if necessary, before assuming his new duties. Forbes replied that his ongoing research work and administrative arrangements would preclude his coming to Champaign until January 1, 1885, and after negotiations he accepted the offer. In June, Peabody reported to the trustees: "The University may surely be congratulated upon the acquisition of so valuable an instructor and so distinguished a naturalist as Professor Forbes. I am confident that the Trustees and the Faculty will support him in every reasonable way."

In fact, the trustees came through splendidly, eventually supporting a total salary of $3,000 for Forbes's teaching, and his research—half again larger than Peabody's initial offer and any full professor's salary at Champaign—and congratulating Forbes for having been elected to the chair of zoology and entomology.

That summer, with Peabody's approval, Forbes put a temporary hold on completing his appointment at Champaign until September. Forbes needed some time to discuss his directorship and the fate of the state laboratory with authorities at Normal without the *fait accompli* of his move to Champaign and the publicity associated with it. Peabody assured him: "Our people are disposed to be quiet and to do nothing that should mar the present pleasant prospects as to your work." The board of education at Normal responded by granting Forbes a leave of absence without pay, allowing him to proceed unhindered with his plans. Their eventual approval for the movement of the laboratory to Champaign, simultaneously freeing up substantial much-needed classroom space at Normal, was unanimous.[17]

"A queer life we are living here," Forbes wrote Nettie in November 1884. Clara and their four children had been living in their new house in Urbana almost two months, while Forbes stayed with his in-laws and completed his work at Normal, but "I keep the railroad from here to Champaign pretty warm," he added. He had been welcomed heartily at the university but dreaded the first year "because I shall be dissatisfied with myself, whatever others may think of me, but after that I shall begin to do my purposes justice."

After the completion of remodeling work in late 1884, Forbes's chief purpose was to move both the state entomologist's office and the state laboratory into University Hall on the Champaign campus. The office would occupy a room on the first floor, the lab three newly subdivided rooms in the basement, including a smaller well-furnished laboratory accommodating 15–20 students. Finally, Forbes requested use of the first-floor tower room for his classes. Along with his equipment and microscopes, Forbes lugged to Champaign his collection of tens of thousands of specimens of fungi, fish, reptiles, insects, and other animals, as well as a library of 1,207 books, and 3,856 pamphlets, papers, and periodicals. In the midst of this academic safari, Forbes received a letter from an old friend, Edmund J. James, professor of public finance at the University of Pennsylvania, saying he had recommended Forbes for the presidency of Iowa Agricultural College. Surprised and amused, Forbes had as much interest now in Iowa as he had about the moon.

At their spring 1885 session the Illinois General Assembly formally approved, by legislative enactment, the transfer of the state entomologist's office and the state laboratory. It also approved a name change from Illinois Industrial to the University of Illinois. Proponents of the change argued that the word "industrial" was misleading and that a less restrictive name was needed. They had no intention—as suspicious opponents charged—of changing the university's overall dedication to "such branches of learning as are related to agriculture and the mechanic arts and military tactics, without excluding the scientific and classical studies."

Forbes brought to the University of Illinois wide distinction in several areas of biological science, and he arrived at an opportune time when the regent and board of trustees were intent on developing and strengthening the sciences. The trustees clearly stated their support for Forbes and his work, saying that the university would do whatever was necessary so that the state laboratory "may enjoy a commodious and perpetual home within, and the generous cooperation of an institution founded and maintained for the promotion of scientific research and the dissemination of practical knowledge." Forbes could hardly have asked for anything more.[18]

FORBES AND

THE RISE OF

AMERICAN

ECOLOGY

CHAPTER 5

State Entomologist and Professor, 1882–95

T he soon-to-be-named University of Illinois at Champaign-Urbana, whose College of Natural Sciences faculty Forbes joined in early 1885, showed more than a few rough edges. Situated on a flat, expansive prairie, its main links to the outside were by railroad and telegraph. The adjoining sister towns boasted just under 10,000 inhabitants and wooden sidewalks but suffered unplanked roads that in wet season churned into a bottomless quagmire. University enrollment in 1884 tallied only 362 mostly rural youths, with one-quarter of them toiling away in a much-needed preparatory department. They were taught by a modestly paid faculty of about thirty, of whom almost 60 percent were full professors.

In boldly bringing along both the State Laboratory of Natural History and the State Entomologist's Office from Illinois Normal University to Champaign County, Forbes did so with the understanding that the work of both "was to be merged and managed as one." Eventually he could not discuss the work of one without mentioning the other. Although never a University of Illinois department, the entomologist's office, which Forbes would hold for the next thirty-five years, became "merely a differentiated part of the natural history survey of the state."

Previously lacking state appropriations and property, and fully dependent on the private library and insect collections of the state entomologist, the office now all-at-once prospered. Forbes and his office gained access to the library, collections, quarters, and research assistants of the state laboratory, and the entomologist's office obtained its own appropriations along with those of the lab. Forbes's collections were augmented in 1883 with 150 boxes of insects bought from country physician W. A. Nason of Algonquin, Illinois. Then in September 1885 Forbes learned that Fanine LeBaron of Elgin, daughter of former state entomologist William LeBaron, had given the university a large, excellent insect collection made by her late father. This gift more than tripled the number of representative insect species in Forbes's collections. He reported that all collections, apparatus, books, and facilities under his charge were freely available to his zoology and entomology classes—especially those for postgraduate and special students. In this fashion Forbes skillfully integrated his research, service, and teaching for the benefit of students at the university.

As the new professor of zoology and entomology Forbes taught three courses: a one-year sophomore zoology course, a practical entomology course, and a one-term introductory general zoology course for "literary students." He also offered laboratory work in histology and embryology, as additions to the usual comparative anatomy and systematic zoology. His sparkling new zoological laboratory, filled with modern furniture, equipment, materials, and microscopes, was now, he said, "conveniently fitted with everything . . . to accommodate thirty-six students at a time."

Insects were a problem from the start for early settlers on the frontier. Native insect species switched from eating wild plants to large stands of cultivated crops, and settlers unknowingly brought along nonnative insects from afar, even as other species voluntarily migrated in. Before 1870 naturalists, physicians, and self-trained entomologists in the midwest had collected, named, identified, studied, and even published about many insect species. Yet they lacked adequate knowledge for controlling insect depredations. This practical know-how was soon championed, for example, by Illinois state entomologists Benjamin Walsh and later Cyrus Thomas. After Forbes's appointment to that office in 1882, Thomas wrote:

> You ask me for suggestions. . . . The *chief work* should be confined mainly to the few, not more than a dozen, notably injurious species. The Plum Curculio, Hessian Fly, Chinch Bug, Cabbage Worm, Codling Moth and a few others are the insects to study. . . . The fact is we scarcely know the entire life history of a single species. . . . *Bear in mind* in writing your report, that *it is for the farmers* and not for scientists, hence technical language should be dropped. . . . In this work you should get right down to the practical.

Up until 1888, Illinois, Missouri, and New York were the only states supporting sustainable work in economic entomology. Illinois farmers, saddled with debt during the 1870s, suffered falling crop prices on through the 1880s and drastic declines through the mid-1890s. Coupled with drought and marauding insects, they were triply besieged. The chinch bug, for example, a tiny dark-colored insect attacking cultivated and wild grasses, corn, and grains, was estimated to have caused $350 million damage in the United States during the half-century before 1900. In Illinois alone it was responsible for tens of millions of dollars in damage, yet it was only one of hundreds of insect species attacking similar crops during Forbes's time. The chinch bug received a good portion of Forbes's attention, coinciding with periodic outbreaks of the pest during his tenure. Along with work on other corn, cereal, and forage insects, fruit insects, and his difficult work on insect diseases, all during 1883–92, these represented the most diverse and sustained economic entomological contributions he made during his long career. In large part they led to his election as president of the Association of Economic Entomologists in 1893–94 and brought him acclaim as "the leader among economic entomologists in America."[1]

Soon after Forbes embarked on economic entomology he wondered whether the information he provided concerning injurious insects was actually reaching the farmer, whether he read it, understood it, and actually applied it to his farm practice. The remarkable way that Forbes dealt with this is a revealing measure of the man and a significant reason for his legacy of "science in the public service."

At a farmers' institute (a public lecture and discussion) very early in his tenure (1883), Forbes discussed how the economic entomologist went about his business. He began, Forbes said, by learning what damage was preventable and how this damage might be prevented at least cost.

The relevant science dealt with insects, plants, and farming, and the interactions among them. A thorough knowledge of insect life histories (reproduction, growth, and survival) served as a foundation for the work but needed attention. This was supported with data on enemies, friends, disease, weather, and season. Learning which factors favored or hindered the insect(s) of interest were next for knowing when to allow natural causes to check insect outbreaks, or perhaps even stopping farmers from destroying, Forbes said, "our best [insect] friends under the supposition that they are the authors of the mischief which they are exerting themselves to prevent." Finally, the entomologist had to understand how farming practices assisted or hindered the pests and which additional means might be used to destroy them.

Here, repeated scientific experiments were necessary: the first should be small and reasonably controlled; the second, larger and conducted under varying field conditions. Observe, record, generalize, experiment, and verify— these were the principal scientific steps required, Forbes noted, a considerable improvement over earlier entomologists who were satisfied merely to make inferences from their observations and apply these directly to entomological problems.

With mounting interest the audience could clearly understand the difficulties with this work, the way he combined practical with pure science, the skill and energy needed for its design and execution, the equipment, facilities, and opportunities for operations, and the dedication and patience needed for its success. Most telling, he admitted ignorance of certain key facts, told them *they* could materially help fill in needed pieces of the puzzle, and asked for intelligent criticism of his work. In this fashion Forbes gained the admiration and trust of farmers.

He then gave them a vivid perspective of the corn plant and its enemies: "Every part of the plant, at every stage of its growth, from its cradle in the earth to its grave in the granary, is regularly taxed to support a ring of plunderers who have fastened themselves upon it, draining its life and appropriating its substances." Specific insects attacked not only each part of the corn plant but also stored corn and even the ground meal. Forbes could cover adequately only a few examples on his list of insect plunderers of corn that night in 1883.

Then there were the hundreds of other insects attacking cereal and forage crops, fruits, vegetables, trees, and ornamental plants, an impressive number of which occupied Forbes and his assistants in the laboratory, the insectary (insect lab), and in the field. They tried to identify, learn the habits, and control a legion of these tireless competitors, who wanted, Forbes remarked, "the same things at the same time that man does."

In short, for comfort we need to consider a selected number of injurious insects, giving priority to those dozen or so species appearing in Forbes's publications that he considered "more important," half of which still cause moderate to serious damage today in Illinois to fruit as well as cereal or forage crops, and by larvae or young alone, or by one of these stages and the adult insect. Table 2 lists a dozen such insect species attacking crops in Illinois during the half century after 1880.

Most of the cereal or forage insects were controlled by agricultural practices; fruit insects required chemical control supported by wise farming practices. Forbes also did pioneering work in biological control, that is, using disease, parasitic, or preying organisms to help keep injurious insects in line.

Agricultural practices made conditions unfavorable for insect pests, and increased the growth, vigor, and survival of the plants. Forbes's guiding motive was to raise farmers' awareness for improving these practices for both their

Table 2.

Some More Important Injurious Insects

Cereal or forage	
Chinch bug	*Blissus leucopterus* (Say)
Corn billbug	*Sphenophorus* sp.
Corn root aphid	*Anuraphis maidiradicus* (Forbes)
Hessian fly	*Mayetiola destructor* (Say)
Wheat-stem maggot	*Meromyza americana* Fitch
Wheat jointworm	*Tetramesa tritici* (Riley)
White grub	*Phyllophaga* sp.

Fruit	
Cankerworm	*Paleacrita vernata* (Peck)
Codling moth	*Cydia pomonella* (L.)
Fruit bark beetle	*Scolytus rugulosus* (Müller)
Plum curculio	*Conotrachelus nenuphar* (Herbst)
San José scale	*Quadraspidiotus perniciosus* (Comstock)

Scientific names of insects after Davidson and Lyon (1987).

own and their neighbors' benefit. It made no sense, he asserted, for some farmers to "breed hordes" of chinch bugs, Hessian flies, or plum curculios in their fields, and then send the beasts over to their more enlightened neighbor's land.

Fundamental to successful farming was crop rotation: corn to nongrass like flax, clover, or buckwheat; wheat to legumes—both would cut down insect injuries to corn and wheat on the return cycle. Legumes provided dense shade and extra moisture unfavorable to pests, and were allowed to appear naturally or were scattered intentionally among the main crop.

Plowing or harrowing of soil and fall plant debris killed insects outright, or it buried larvae so deeply, Forbes said, that they were unable to reach the surface after leaving the larval stage.

The timing of planting also offered advantages. Early sowing of grasses (except corn) and grains liable to insect attack advanced the crop out ahead of the insects. Late fall sowing of winter wheat, for example, protected young plants from early fall broods of Hessian flies.

The remarkably simple—but often ignored—habit of clean culture paid rich dividends. Removal of plant debris, rotten fruit, diseased wood, tall grass, and weeds from fields, fence lines, and woodland edges meant fewer hiding places and fewer extra foods for a wide variety of insect pests.

Lastly, Forbes recommended that extra fertilizer be added before and after planting, having good evidence that this practice helped plants resist or cancel insect injury.[2]

In one form or another these themes appeared repeatedly in Forbes's writing and addresses. He stayed square on message: a lesson in what was applied ecology covering the interactions among insects, plants, and humans. And in doing so he snatched the advantage from those who grumbled that "nothing could be done" to combat insect pests; that they were like the weather and other God-given natural events; that these farm practices had been tried (often at the *height* of insect infestation) and "didn't work"; that "how could anyone tell a maggot from a grub anyway?" (he clearly detailed this); and that "my neighbor found (this or that) insect on his rotting cabbage, so *it* must have been the culprit." Coincidence, he preached, does not prove cause. Hence Forbes praised public virtue and threw his wholehearted support to those astute farm families who used their curiosity and imagination in learning more about connections between the farming enterprise and the natural world, and who considered setbacks at the hands of insects as only "preliminary skirmishes" in a lifelong campaign. After all, the insects had been around far longer than humans, were often "better organized," and certainly more abundant. He also emphasized the benefits that humans received from insects: as scavengers, as pollinators of fruit, vegetables, clover, alfalfa and ornamental plants, as makers of silk and honey, as destroyers of injurious insects, and as food for birds and fish. In short, Forbes put things in perspective, and boosted the farmers' courage and optimism to stay the course.

During his first hectic eighteen months as state entomologist, Forbes tangled mostly with insect pests of corn, wheat, oats, and hay. He often used military metaphors to describe his work: for example, as "the general advance of a well-supported skirmish line, covering, by way of reconnaissance, ground on which we expect to do some heavy fighting hereafter," namely by attack in force at strategically important points, and he wrote, by "a union, under one management, of various related kinds of public natural history work."

In order to increase the flow of information on injurious insects to the public, Forbes boldly challenged the state legislature to appropriate adequate funding for his entomological reports and monographs by refraining to present his formal annual report as state entomologist for 1885 and leaving it to the legislature to provide for his publications hereafter. They did so beginning the following year. His cultivation of the legislature for almost a decade coupled with his productive use of state support was now a potent force for the future success of all his programs.[3]

In June 1885 Forbes began testing the efficacy of insecticides for codling moth control in apple orchards. Two standard insecticides were available: arsenicals as contact poisons for gnawing insects, and kerosene, soap, and water emulsions for sucking insects. Before Forbes's work, reports on arsenicals were very general and vague, and no one had employed experimental work.

Some detail on this work will show how Forbes carried out his field experiments and the conservative recommendations he made concerning the use of control chemicals potentially dangerous to the environment and public health, traits which his colleagues came to admire and use as standards for their own work.

Forbes tested the two arsenicals, Paris green and London purple, mixing them with water and supplying two gallons of fine spray to each of six experimental trees several times between June and September. Six other trees were untreated and served as controls. His assistant picked up fallen apples and all intact ripened apples, cutting them in two, examining them for codling moth larvae and recording injuries in more than 16,500 apples. Referring to picked ripened apples, Forbes noted that "Paris green would save at a probable expense of ten cents per tree, 70 percent of the apples usually damaged by codling moths, and London purple would save 20 percent.

But there were additional concerns, Forbes cautioned. Experimental trees had scorched foliage, loss of leaves, and discolored fruit. Also, apples picked in September and sprayed recently showed arsenic residues averaging 0.9 mg, despite exposure to rain. Eating seventy-four apples over time would have delivered a poisonous dose to a human. September spraying could not be permitted.

He also had to take into account that codling moths had two broods of young each year in central Illinois, the second and larger one as late as September. Spraying them in September would also be dangerous.

Experiments with arsenicals repeated soon after at four other state experiment stations, supported under the new Hatch Act and in response to Forbes's lead, also showed plant damage, even at weaker arsenical mixtures. Charles V. Riley, chief of the Division of Entomology, USDA, suggested correctly that plant damage was directly related to the percentage of water-soluble arsenic. Dry arsenicals appeared better.

Forbes also showed that spraying should be done less frequently (even once or twice was enough to obtain good results), and that spraying should be done after blossom fall and fruit formation to protect bees. He estimated savings of $1.5 million annually in the average Illinois apple crop if these improved methods were followed.[4]

The Hatch Act, signed into law by President Grover Cleveland on March 2, 1887, established an agricultural experiment station at each of the land-grant universities and soon provided $15,000 annually to each state for doing the necessary work. Forbes and Thomas Burrill believed that such stations would be "of vast service in the promotion of theoretical and practical agriculture." Their ultimate goal was for an enlightened agriculture, one guided more by science than by a lackluster routine handed down from father to son. They wrote: "So as the processes of agriculture become scientific and rational, rather than empirical and traditional, the value of experiment and investigation becomes recognized, and new knowledge is not only tolerated, but is more and more sought for by practical men."

The Illinois experiment station opened in the spring of 1888 with an office in the chemistry building, and an eager staff of four. Research already underway in agronomy, meat and milk production, horticulture, and insect control, made up the initial agenda. Over the next two years station workers carried out more than a hundred studies and published a dozen bulletins, including those of Forbes, who had volunteered his services to the station without pay as consulting entomologist.

The Hatch Act had wide-reaching consequences, prompting Forbes to examine the work of both the state laboratory and entomological office in reference to the new experiment station. The overall purpose of experiment station work was clearly economic; consequently, Forbes wrote, "its scientific work must naturally be regulated with close reference to practical results." Yet many species of plants, insects, and other animals, although of no immediate economic interest, were intimately related to economic species in their classification, habitat, food and feeding, and other characteristics. Both groups of species interacted in the natural world. And so he believed a sizable knowledge of both "is very helpful, and often indispensable for the solution of merely economic problems." In fact, more purely scientific work at both the state laboratory and the entomologist's office could well provide a broader and stronger foundation on which the economic work of the experiment station could build.

The state of Illinois had supported "fairly well" both scientific and economic entomology for some years. Forbes's strategy had been to run practical field experiments likely to provide economic answers. But these were highly expensive and time-consuming, and they provided neither adequate control of conditions nor easy observation of insects. For now he would put them aside in favor of other work for the experiment station, such as identifying insects of economic interest, studying life histories (reproduction, growth, and survival), and doing office experiments on insect lives and habits perhaps needing field verification. To do his experimental work now, though, Forbes needed a special entomological laboratory, or insectary. This would allow breeding of insects, rearing of their food plants, and more accurate experiments—all under more tightly controlled conditions and continuous observation. Hence he asked the state legislature for $1,000 for a suitable building. His request was granted in the 1889–90 biennial budget.

With his insectary in place, Forbes got a jump-start in his studies on some of the more difficult species. These covered the life histories, feeding habits, and injuries to plants of half of the more important injurious insect species mentioned earlier: the European bark beetle, chinch bug, corn root aphid, plum curculio, Hessian fly, and white grubs. These species appeared in nine of Forbes's informative bulletins from the experiment station during the next ten years, stressing crop injuries and remedial economic measures for the practical farmer and horticulturist, and secondarily, descriptions, lives, and habits of the insects for the more scientifically curious.

Still, some effects of the Hatch Act brought mixed blessings. For example, of the forty-three states with experiment stations, thirty-one were doing some entomological work, but not all researchers involved were trained entomologists. Nor was an adequate number of entomological assistants available after 1888; consequently, it was a seller's market and would be for another ten years until university programs had increased the supply of entomology graduates. Forbes felt the effect of this shortage when he lost several valuable men because of insufficient pay for entomological assistants at Illinois, who were stuck at annual salaries of $600–900.

Furthermore, in his progress report on economic entomology in the United States for the U.S. Office of Experiment Stations, USDA, in 1891, Forbes pointed out the one-sided character of much experimental work during the previous year. Most of the work was horticultural (fruits, vegetables, and ornamental plants), with emphasis on chemical and biological control methods. Absent, but needed, was work on farm practices and management for preventative and remedial insect control. This was a high priority field where several stations might contribute in an important way.[5]

During the time under discussion, Forbes was in his middle years. He stood 5 feet 10 inches tall and had an erect carriage. Some saw in his physique the typical Scot from Aberdeen; others, the cavalry officer he had once been; and still others, the urbane professor who ably dominated a room with his presence and fine oration. A light sleeper, he rose at 5:30 after five hours rest, breakfasted, and then wrote and read for an hour or two before arriving at his office at 9 sharp. Still a prodigious reader, he devoured philosophy, ethics, pedagogy, metaphysics, genetics, history, politics, methods in science, poetry, French fiction, and music, also playing the organ for years before pianos became popular. Reading and study were accompanied by his cigar or pipe, habits from army days and continued well into his sixties. He also chewed tobacco for many years. He presided as first president of the university golf club, and regularly biked, saying it added ten years to his life. Forbes claimed he was the best rider in his cavalry regiment and insisted that all family members learn to ride. Clara drove the family surrey, the popular high cart, and the sleigh, and "gave more credit," said youngest daughter Ethel, "to the horse than to any person for bringing up the family."

To his young children, "the little folk," he was "Papa" who read to their mother in the evening, who took frequent train trips far away, and who came home with little gifts tucked away in his magic valise. And it was he who supplied the family with a horse, a cow, a Newfoundland dog, chickens, rabbits, and an "unlimited supply of cats." He expected much of his children as he did of himself. Where he believed he fell short was in not always holding his temper with them, in not more closely guiding each child toward appropriate accomplishments useful for their future, and in not giving better guidance in religious matters. Perhaps the more liberal thought he and Clara shared was too diffuse, too general for their likes when they had a limited amount of other examples to choose from freely.

At the time, in the tricky passage from an agrarian to an industrial society, strong paradoxes emerged in the values most Americans shared. Material wealth steadily increased, but so did corruption, injustice, and ethical shabbiness. Progress in technology took the breath away, but complexity and "bigness" dwarfed the individual. Farmers, for example, who made up one-third the population of Illinois, felt that banks, markets, and railroads constantly pulled their strings, and found themselves working harder and harder within an increasingly commercialized agriculture while having less and less control over their own production. During his long career Forbes worked hard to get farmers to unite in combating injurious insects, but found it an exceedingly formidable task. In business, in industry, in government, and in growing cities, problems multiplied and confusion abounded. As people sought order and integration, the country churned and became more diverse.

As a thorough rationalist Forbes believed that the human species was a product of both its heredity and its environment, and that it was occasionally a victim of forces beyond its control. But he believed, too, that the human species and the world were not finished evolutionary products. Humans could think, reflect, investigate, and invent; that was their genius. They could, therefore, act to improve themselves and the world. Forbes wrote:

> The cure for the evils of progress is more progress; as we change our environment we must also change ourselves. . . . As we are drawn inevitably and irresistibly, year by year, into closer bonds of social and industrial companionship,

we are bound to become socially and industrially more companionable; we must consent to restrictions, in each others' interest, which a generation ago would have been thought intolerable; we must volunteer habitually mutual services which were once uncalled for.

Concerning his spiritual beliefs, Forbes resolutely held himself up to a high and realistic level of proof for Christian membership, but he fell short and so became an agnostic because:

> One must subscribe to a certain form of belief—and that I do not hold. . . . I must deny myself the right to be called a Christian—not caring to twist the meaning of the word to fit my deficiency. It is scarcely possible, indeed, that I shall ever again become one, for to me my departure from the old way has been a process of slow and almost unconscious growth, both of thought and feeling, and not in any sense a process of degeneration. I have of course suffered innumerable inconveniences as a consequence—mostly in the way of a certain isolation among my friends. . . . All this is inevitable and I long ago accepted it. . . . I do not complain. My life, is in fact, a happy one.[6]

Happiness increased for the family when the last child, Richard, was born in 1887. Stephen and Clara sought out the newly formed Unitarian Society that sponsored a series of lectures in Urbana in 1888. Thereafter, they fully supported liberal religion, holding occasional meetings in the parlor of their Springfield Avenue home, and then organizational meetings for a 1907 establishment of the Unitarian Church in Urbana, of which they were founding members.

Forbes's organizing ability and steadiness of purpose were instrumental in his 1888 election as dean of the College of Natural Sciences. The schools of chemistry and natural history making up the college had a total enrollment of about fifty students. Within only two years enrollment would double, chiefly because of faculty changes and additions to chemistry, supported with funds by the second Morrill Act of 1890. Natural history included the departments of botany (Thomas Burrill presiding), geology (Charles Rolfe), and zoology (Forbes).

Despite serving as the chief administrative officer of a college, a dean at Illinois before 1894 did little more than preside at faculty meetings, had nothing to do with students, and exercised no control over heads of departments. Nor curiously was he ever addressed as dean. All this would change in a few years. Yet similar to other obligations he had assumed, Forbes brought to his first deanship thoroughness and accuracy. As his friend and colleague Theodore Pease would say years later, "He would not trifle even with trifling things; he would touch no piece of work to which he could not give the best that was in him."

In the late 1880s total enrollment at the university numbered almost four hundred, 20 percent of whom were women. Tuition was free and fees for the four-year course amounted to $105. Engineering remained the largest college including more than half the students in the university. Natural science held its own, since chemists and natural science teachers were notably in demand, but the College of Agriculture was dormant, with fewer than two dozen students. Nevertheless, the soil for recovery had been prepared, and the seeds for an invigorated agricultural science at Illinois were then planted and cultivated by Forbes and Thomas Burrill.[7]

A few years earlier experimental chinch bugs living in some corn plants in Forbes laboratory began dying. Why? he wondered. Young and old alike were affected. He sacrificed some living bugs and found their body fluids swarming with bacteria, but found no bacteria in corn juices they fed on, nor in water he had washed over the bugs before examination. Next he took additional bugs, examined various body parts one-by-one, and found bacteria only in the posterior portion of the intestine, specifically the whorl of caeca or Malpighian tubules in that region.

Three weeks later he revisited the cornfield home of his chinch bugs, this time finding bugs much less abundant than before. Forbes found many dead bugs, both young and old, behind sheaths of corn, and noticed fewer older living insects than in other fields 1 to 2 km (1 mi.) distant. Both living and dead bugs collected during this revisit carried abundant bacteria seen easily with his immersion microscope lenses. Only available in the last few years, these new "homogeneous" lenses could be lowered directly into specimen fluids under his microscope for up-close viewing. Thomas Burrill decided this bacterial species was new to science and named it *Micrococcus insectorum*.

As the new state entomologist Forbes found himself consumed with the most serious chinch bug explosion in Illinois since the early 1870s, and what would ultimately be the worst, longest, and most widespread chinch bug infestation in Illinois history to that time. Corn and all grains, but particularly wheat, were devastated. In addition to the new *M. insectorum*, Forbes found other dying bugs infected with the fungus *Entomophthora* as well; he believed that both were potential natural checks on chinch bug populations. Coincidentally he also stumbled on tent caterpillars in orchards dying with fungal infections and silkworms in Burrill's laboratory dying with bacterial infections.

What with his new immersion lenses, new laboratory media for culturing microorganisms, and his familiarity with the recent celebrated European work on "germs of contagious disease," Forbes felt "the time [was] ripe for pushing our studies into this field." He took note of Louis Pasteur, whose research during 1865–70 with pebrine (black-spot) and especially flacherie (dysentery) diseases of the silkworm had literally saved the French silk industry.

His immediate task now was to carry out laborious culture and infection experiments to gain experience with a specific microbial disease. He chose the bacterial infection of the introduced European cabbage worm *(Pieris rapae)*, cabbage worm white plague he called it, with symptoms similar to Pasteur's flacherie in the silkworm: green to progressive milky-white to gray body color and increasing loss of activity. During September 1883 the disease appeared in patchy fashion in central and northern Illinois, and in Michigan. Eventually the disease ran rampant that fall in east-central Illinois; it had not been present the previous year. Forbes identified the culprit bacteria as a *Micrococcus*—in all probability *M. bombycis,* a silkworm bacteria. He also found what appeared to be the same bacteria in the apple caterpillar *(Datana ministra),* with similar disease symptoms. Forbes then had some success in conveying the microbe via dead and diseased caterpillars into the field and infecting other healthy larvae living on farm crops.

Accordingly, with proper controls in pathbreaking work, Forbes showed the transmission of disease bacteria to several species of moth caterpillars and to white grubs (beetle larvae), the principal effect of the disease being erosion of the epithelial layer of the intestine. Burrill served as advisor for the work, writing Forbes that he "had done well for [it being his] first time in this field," especially since bacterial diseases were less understood and more difficult to work with than most other insect diseases. Charles Riley of the USDA wrote that Forbes's work was "of great scientific importance" but doubted whether farmers would soon benefit.

In early 1887 Forbes had the opportunity to review the current status of contagious insect diseases. He had served as president of the Cambridge Entomological Club during 1886–87, and addressed the club at Harvard on this topic as retiring president. The club served as a good forum long having had a remarkably wide interest in all biological aspects regarding insects and having kept in touch with entomologists and their work across the country. Forbes said he had been unable to transmit pebrine disease of silkworm caused by a sporozoa protozoan to a variety of caterpillars. Muscardine, or fungal infections, however, the longest known of insect diseases, had better promise for insect control. Fungi could be readily obtained in pure culture with several liquid or solid media, or roughly transferred from one group of bugs to another by use of a simple "contagion box (or a bug sick bay)."[8]

In 1888 Forbes first came up with the fungus *Beauvaria globulifera,* which caused widespread mortality of chinch bugs in Clinton County, Illinois. Shortly thereafter it was identified from Minnesota, Iowa, Ohio, and Kansas.

First used for insect control by O. Lugger in Minnesota (1888), this fungus underwent exhaustive pioneering use by F. H. Snow in Kansas (of whose reports, however, Forbes was particularly critical), and by Forbes during the 1890s. Their work stimulated other American researchers' interest in the possible control of insect pests by selected microbes.

In Illinois Forbes supplied both infected bugs as well as cultured fungus to several thousand eager farmers in more than 600 towns who volunteered to try the white muscardine fungal disease against chinch bugs. They were asked to follow easy suggestions for use, set up a simple contagion box of their own so that infected bugs could be repeatedly added to their fields, and to report back regarding their observations. Despite some success with the introduced disease in some areas where large bug populations had not yet encountered naturally occurring fungi, in general Forbes found the fungus present naturally to such an extent that human spreading of additional spores or infected animals seemed unnecessary. It was clear also that high humidity and wet conditions assisted spread of the fungus in either event. Work stopped in other states as well, as the chinch bug explosion of the 1880s slid into more scattered outbreaks of the 1890s, followed by a welcome twelve-year lull.

Soon entomologists explained that complete eradication of an insect pest, which the public expected, was not a realistic goal. Instead, reasonable reduction of a pest to a lower population level was a better tactic, a goal that Forbes had embraced and worked for. Lower population levels of insects meant a commensurate decrease in potential insect damage, and if humans could initiate or speed up natural control by enemies of insects in a careful way, so much the better ecologically and economically.

Forbes's limited success with bacterial diseases stemmed from his ignorance at the time about the role of viruses in insect diseases, which would not be demonstrated until the 1920s. Paillot in 1930 believed that viruses, in fact, were the probable initiators of Pasteur's flacherie disease of silkworms with bacteria acting as secondary invaders complicating or even strengthening the disease. Viral presence in these infections sometimes lay in visible crystal-like bodies, or polyhedra, that were occasionally recognized during the nineteenth century. Forbes himself identified a "polyhedrosis" disease of the armyworm (*Leucania unipuncta*), believing it to be a potential natural check on armyworm populations.

The legacy of Forbes's pioneer work in insect diseases for use in biological control of insect pests would appear later in the twentieth century, specifically as a foundation stone for integrated pest management (IPM). A good example is the use of bacterial and especially fungal insecticides for control of gypsy moth populations, the latter of which helped dramatically reduce gypsy moth numbers in heavily infected northeast states beginning in the 1980s.[9]

CHAPTER 6

Ecology at the Center, 1891–1917

I n 1891, while Forbes labored with economic entomology and his other
manifold duties, the University of Illinois lay on the brink of a significant
transformation. Early in September, Regent Selim Peabody resigned, and
the board of trustees appointed Thomas Burrill as acting regent. Burrill had
been an advisor and supporter of Peabody and the board, and few suspected
from Burrill anything more than a caretaker's role as acting regent. But this
was not to be.

Burrill's first action that fall was to convey more power to the faculty in setting university policy, including a provision for faculty to elect deans of the
colleges. Burrill also enlarged the curriculum by 45 percent in courses offered
(mostly advanced ones), and with the board he successfully obtained from the
state a third of a million dollars for the 1893–95 biennium. These funds plus
federal funds from the second Morrill Act of 1890 did wonders for university
instruction.

Burrill finished his three-year stint as acting regent by raising faculty
salaries, increasing the faculty, formalizing a graduate school, and firming up
the sciences and engineering departments. The 50 percent increase in student
enrollment at Illinois between 1891 and 1894 was the most rapid growth of any
state university in the country.

Burrill's presence as acting regent encouraged those faculty who believed
that science education at Illinois needed upgrading. Consequently, in 1892 the
old College of Literature and Science was split into separate colleges. The re-
organized College of Science was comprised of four groups: chemistry, physical science and mathematics, natural sciences, and mental sciences. The natural

science group included as before the departments of botany, geology, and zoology, headed by Burrill, Rolfe, and Forbes. Enrollment in this group increased to sixty-two in 1893–94, almost double that in 1891; one-third of the
students were women. As dean of the College of Science the faculty designated Stephen Forbes.

Former regent Peabody had received earlier in 1891 a state appropriation of
$70,000 for a new natural history building. Designed by Nathan C. Ricker,
the first American graduate of a university architecture program, and built in
Second Empire style, this handsome red brick, three-story building became
home to the natural sciences, whose faculty moved in during the fall of 1892.
Four rooms on the second floor were provided for Forbes and the State Laboratory of Natural History that moved in six months later. Forbes was pleased
to secure his brother, Henry, as business agent and librarian for the laboratory,
a career switch for the colonel after twenty years as principal and teacher. The
dedication of the new building in mid-November with 1,200 people attending
featured Forbes's good friend and zoological colleague Dr. David Starr Jordan,
president of Stanford University, as principal speaker on "Science and the
American College," and that evening at a reception, on "Agassiz and His Influence." Thomas Burrill spoke on "Development of the Natural History Departments." At the dedication Jordan said:

> We are here to raise a mile-stone on the road of education. We are here to cele-
> brate the emancipation of science from the bonds which held it down twenty
> years ago. Then was a time of depression in which science was almost entirely
> omitted from the curriculum of the best colleges. . . . The first upward step for
> the State university was the adoption of the elective system—the substitution
> of advanced science for elementary studies in other departments. I congratulate
> the State of Illinois that its University has all its [Natural Science] departments
> in one place, for the student of each department will be the better for rubbing
> up against men who are not in his line of study.[1]

Three years earlier, following the formation of most state agricultural experiment stations, entomologists at the stations had felt a strong need for
their own professional organization. Charles V. Riley, USDA chief entomologist, urged this step and was duly elected president of the new Association of
Economic Entomologists, organized in Toronto during the summer of 1889.
Two dozen leading American and Canadian workers—including Forbes and
his protégés Charles Weed (Ohio Experiment Station) and Harrison Garman
(Kentucky Experiment Station)—were elected charter members. With modest
yet steady growth, the association brought together entomologists from both
federal and state service, and for a quarter-century remained the only association of its kind in the world. By 1907, when Forbes was president-elect for the

second time, membership would increase tenfold, and the association would begin publishing its own *Journal of Economic Entomology*.

Forbes first took office as president in 1893. He gave his inaugural address August 14 in Madison, Wisconsin, in conjunction with summer meetings of the American Association for the Advancement of Science. He celebrated the enthusiasm and dedication of his colleagues, gave an overview of relevant studies during the past year, and presented a constructive appraisal of their work. Of central concern was to ensure that more of the knowledge they gathered be put to use by the farmer, gardener, or fruit-grower. This could best be achieved not by expecting practical agriculturists to become apprentice entomologists, but rather by providing them with descriptions and summaries of insect injuries to specific crops and their treatment—a tune he had sung before.

Most troublesome was what to do about those occasional farmers who made little or no effort to use preventative or remedial treatments for injurious insects. Forbes suggested the possibility of legal inspections and enforceable penalties for farmers "who breed insects by the bushel" that then devoured their own crops as well as those of nearby industrious and intelligent farmers. In conclusion, he touched on a perennial problem of natural resource work—the fact that state and other political boundaries are often artificial, while scientific problems range indifferently across these boundaries leading to duplication of effort and lack of communication. Groups of volunteer entomologists willing to work cooperatively on wide-ranging projects of general interest, he felt, might be a solution.

Ten days later the International Botanical Congress took its turn with meetings in Madison. Unceremoniously, botanists doing ecologically oriented work in what was then a specialized division of physiology, suggested changing the word "oecology" to "ecology" (and from "oekologie," after Haeckel, as used and practiced by a number of German botanists and a few zoologists since 1885). Picking up on this topic later that summer, J. S. Burdon-Sanderson, M.D., professor of physiology at Oxford, and new president of the British Association for the Advancement of Science, spoke glowingly of ecology as now being a legitimate division of biology "with equal distinctiveness" from physiology and morphology, and "by far the most attractive." Scientific interest in ecology now spread more quickly for the first time, and despite detractors, on balance prospered during the 1890s with botanists leading the way and urging zoologists to come along.

Forbes had brought an unerring ecological viewpoint to every biological problem he had addressed since 1875. With keen insight, in both his biennial report as state entomologist and his report on aquatic biology work with the

state laboratory for 1893–94, he gave his imprimatur to ecology as a science—
one of the earliest zoologists to do so. For the entomological side, he wrote:

> There is another division of biological science, little known to the general pub-
> lic by its name as yet, and but lately distinguished as a separate subject, which is
> now commonly called *oecology*. It is the science of the relations of living animals
> and plants to each other as living things and to their surroundings generally. . . .
> The whole Darwinian doctrine belongs to it on the one hand, and all agricul-
> ture belongs to it on the other. . . . Economic entomology is, of course, a divi-
> sion of this science of oecology and may be exactly defined as entomological
> oecology applied to economic interests. . . . In practice, indeed, it is to the sys-
> tem of living nature as affecting insects, and through them as affecting man,
> that he must give his principal attention.
>
> The study of oecology is thus to the economic entomologist what the study
> of physiology is to the physician. Human interference with the natural order of
> plant and animal life gives rise to reactions which correspond closely to those
> of bodily disease. . . . As the old medicine was chiefly a matter of drugs and
> empirical dosage, so economic entomology was not long ago almost wholly a
> matter of insecticides and mechanical means of destruction. The new economic
> entomology . . . like modern medicine, seeks diligently, first, to avoid all unnec-
> essary disturbances of the normal play of life, and second, to direct the powers
> of nature herself, so far as possible, to the correction of such disorders as are
> nevertheless likely to arise.[2]

The second goal was a worthy one, yet, not surprisingly, it was one that
Forbes rarely reached. He did, though, contribute importantly to a rational inspection and ecologically friendly control system in Illinois and indirectly the
entire United States for injurious insects and plant diseases. To illustrate this
point, consider the San José scale, a scourge of fruit, nut, and other deciduous
trees and shrubs during the last quarter of the nineteenth century. Many felt it
to be potentially more damaging to fruits than any other insect in the United
States. People were rightly scared.

During the summer of 1880 John H. Comstock, then USDA chief entomologist and a close colleague of Forbes, discovered the insect in the Santa
Clara Valley of California exploding in fruit orchards on bark and leaves,
often leaving fruit unmarketable. Its handiwork is a grayish layer of .05–2 mm
(< in.) scales under which the insect lives sucking sap and killing the tree
within 2–3 years. Crawling licelike young are transported by birds or wind,
and only winged males emerge for reproduction. By careful fieldwork Forbes
estimated that a single female in central Illinois could potentially produce 33
million descendants.

A probable import from China in 1870, the scale moved eastward in nursery stock and by 1893 had spread into several eastern and midwestern states. By 1898 Canada and several European countries had quarantined entrance of U.S. plants and fruits to protect against scale, serving as a strong impetus for the United States to protect itself in turn against import of injurious insects by passing federal quarantine laws in 1912.

Forbes anticipated the scale's arrival in Illinois in the mid-1890s and found some in 1896. He then prepared to combat the scale and "stamp it out where it has already gained a foothold" in orchards and nurseries as urged by his USDA entomological colleagues in Washington, D.C., under their new chief, Leland O. Howard.

Forbes used whale oil soap as a remedial spray for his preliminary work but then tested several other compounds to improve his efficiency. He even tried spreading a Florida fungus at selected locations; it killed scales, but spread too slowly to be effective. In addition, he or the owner cut and burned badly infested stock.

Until 1899 Forbes's work (supported by state funds) required only voluntary compliance of orchard and nursery owners, and fortunately only a few refused entry to their property. Where Forbes had a problem was getting owners to share the costs of his operations. He was convinced that compulsory legislation by the state was necessary to ensure that properties were free of scale, to prohibit orchards and nurseries selling infected stock, and to put legal authority behind sharing of costs to prevent injury to others. Through persistent efforts of Forbes and the State Horticultural Society, these provisions were part of a state act passed in 1899 covering not only the San José scale but "dangerous insects or contagious plant diseases" in general.

By 1901 the scale had turned up and been treated in 243 orchards and seven nurseries, markedly improving the quality and management of the Illinois fruit trade. Forbes's thorough work with scale to this point prompted his election as president of the National Association of Horticultural Inspectors, in which capacity he served for six years.

Not satisfied with damage caused by whale oil soap to more tender fruit trees like peach, Forbes proceeded to put two so-called western U.S. washes (mixtures including lime, sulfur, and salt) to a test on both peach and apple trees with a late winter (dormant) spray in March 1902. In an elegant experiment—including appropriate controls, live and dead scale counts before and after treatment, and simulated rains during five days after treatment to test persistence of toxicity to scales—Forbes found that both washes killed 91–93 percent of scales within five days. Both washes were also harmless to trees and

83 percent cheaper to use than whale oil soap. Forbes now considered the scale easier to control than codling moth, apple and peach borers, and probably easier than the cankerworm.

Two years later Forbes could say that all known scale-infested trees had been treated at least once and, with only three exceptions in southwestern Illinois, had been brought under control by attentive owners. Yet he emphasized that by far, as compared to all infested areas, a greater total area of the state had never become infested by scale at all.[3]

As state entomologist and a research scientist with work supported through the experiment station, Forbes had an obvious interest in a viable College of Agriculture. During his first decade at the University of Illinois, this did not exist. Enrollment in the college averaged about ten per year since his arrival, and the faculty was made up of no more than four professors. This somber predicament was repeated at most other land grant universities, primarily because of the absence of an agricultural science with a tested body of knowledge gleaned from American experience.

After 1895 this changed dramatically at Illinois both because of a growing public interest in agriculture and the needed leadership of the new dean of the College of Agriculture, Eugene Davenport. His fine efforts brought greatly increased state funding for the college and experiment station, increased numbers of students and faculty, added thirty-seven new courses, and a new college building in 1900 with two acres of floor space, including the state entomologist's office and laboratory for Forbes. By the fall of 1905, 430 students of agriculture were taught by 44 faculty members, and the experiment station was inundated by requests for help and study concerning crops, injurious insects, livestock, and soils.

Forbes's work in economic entomology benefited substantially during these years from increased overall support by both the state and federal governments. In the dozen years before 1902 his publications included eleven experiment station bulletins and entomological circulars; in the fifteen years after 1902 that number would more than double. And his papers received quicker, more frequent, and ever-wider distribution. His bulletins, for example, were issued in editions of 50,000 copies and sent via the mailing list and postal frank of the experiment station. An additional 1,000 copies were bound later as part of his biennial report as state entomologist.

Expanding work and cooperation with the experiment station by the U.S. Bureau of Entomology under Chief Leland O. Howard, as well as research funding under the 1906 Adams Act, also helped to support Forbes's injurious

insect work. At the time, Howard recognized Forbes as one of three state entomologists in the United States whose excellent work and reputation clearly
stood out. In fact, Howard said later that "it would have been for the good of
the country" had Forbes succeeded Charles V. Riley as USDA chief entomologist in 1894 instead of himself. Howard had written Forbes then, asking if he
would consider the position. Forbes replied that he much preferred to stay in
Illinois but, Howard said, "I think there is no doubt that he could have had
the position had he wished it."[4]

Justification for Howard's praise can be found in Forbes's work with corn
insects. Over more than three decades from 1882 to 1916, Forbes showed the
longest and most consistent attention to corn insects, publishing some seventy
papers, including twenty written during his thoroughly ambidextrous deanship
years. Then there was his remarkable ability to achieve what he called "conversations" with public audiences centered on economic entomology. Corn, the
most important crop in both Illinois and the nation at large, is worth billions
annually and is frequented by more than two hundred species of insects. Of
these, Forbes singled out about a dozen that were most injurious, and of
these, eight that were capable of the worst damage: corn rootworm, white
grubs, wireworms, chinch bug, corn root aphid, armyworm, grasshoppers, and
cutworms. All of these insects caused injuries to corn roots, stems, or leaves,
especially of young plants, and all remained infamous well into the mid-
twentieth century.

As usual, Forbes stressed what the farmer ought to know and how he might
apply this knowledge to his farm practice, but stressed "that economic entomology is not an exact science." There were too many variables, causes and effects were tricky to pin down, and probabilities were the best he could provide. With these caveats he moved ahead and asked for questions. Or if the
audience was shy, he might recount some of his simple field experiments or
put up lantern slides of injurious species that never failed to excite interest
and went far toward clearing up confusion over common names of insects.
Eventually, in hope of spreading more knowledge of insects, Forbes sent out
named collections of them and an illustrated manual of facts about them to
high schools, figuring the students would tutor the adults.

Early in the twentieth century Forbes fielded more questions about the curious corn root aphid and methods for its control than for any other insect.
About the size of the head of a small pin, these licelike, grey-blue insects suck
sap from corn roots and are moved about to appropriate sites by brown ant
"farmers," who in turn subsist on sticky sweet liquids exuded by the aphids.
The aphids reproduce and overwinter as eggs in the ants' nests. Deep plowing

and disking in late fall, or better, in spring before corn planting, breaks up and scatters ants and aphids better protecting young corn plants from loss of fluids and retarded growth, which are especially bad in droughty years.

Likewise, billbugs often received attention. These dull-black elephant beetles with elongated snouts make short work of young tender corn leaves and stalks, leaving characteristic uniform rows of holes in curled young leaves of plants that survived. Larval beetles also tunnel in corn stalks. Forbes recommended deep plowing and proper crop rotation as protection against a half-dozen billbug species.[5]

Along with his scientific work and his duties as dean of the College of Science, Forbes carried on his teaching between 1892 and 1905. He taught the general zoology major course from his first year at Illinois in 1884 through 1895, as well as advanced zoology from 1886 through 1899, a course in which he had included "ecological studies with a basis in field observations and laboratory experimentation" as one of three main lines of work. Entomology formed part of his course load in the zoology department from 1884 through the academic year 1899–1900, differentiated and labeled variously as general, elementary, advanced, or practical. Lastly, in addition to independent study, he offered an advanced economic entomology research course during 1896–1900 for graduates intending to prepare for experiment station work.

Believing for some time "that entomology was really the larger half of zoology," Forbes felt that it had grown enough to deserve a move toward an independent department at the university. In 1899 he arranged a special four-year entomology curriculum, and the following year six entomology courses were described in the university catalog separate from zoology. Forbes then teamed up with Justus W. Folsom, who had recently earned a doctoral degree from Harvard (1899), and during the next ten years the two men brought to life a full-fledged Department of Entomology at Illinois. Only then did Forbes finally retire from his professorship in zoology. In the intervening years Folsom taught the three elementary and general entomology courses, and they team-taught economic entomology. Both men were available for advanced work and senior thesis. During the last five years of his deanship (1900–1905), Forbes also taught occasional courses such as zoology for teachers and invertebrate zoology.

Then in 1903 his brother, Henry, died. He was 70 years old. Scholar, teacher, guardian angel, friend, comrade, hero, and commanding officer, "his was the finest character I ever knew," said Forbes. For ten years a fixture as librarian-business agent at the state laboratory, and part-time lecturer in Greek at the university, Henry had remained his complicated self: restless and humorous, sensitive and articulate, suffering the frequent shifting moods of depression and elation that had troubled him since his early army days. The colonel said he had often "stood where the shrieking messengers of death were flying thick and fast on every hand, and had been calm and unshaken." But transported back to farming, to education and peacetime pursuits, the sense and round of his life seemed broken, as he predicted earlier that it would be. He served his country and was always giving to family and friends, to whom it seemed he had only a fragment left for himself, and it proved not nearly enough.

For several years Forbes himself had been seeking relief from executive duties. In short, he wanted more time for his aquatic biology research, and for incomplete manuscripts that needed publishing. By 1905 he had been dean for seventeen years, long enough to know the position now offered him nothing more he deemed important. (Later he would admit it had quickly provided his family with position in the university community.) When he had tested the water about resigning his deanship in early 1905, President Edmund James said he would back Forbes's request, that Forbes and Burrill were the only Illinois faculty members "who are widely and generally known outside the University," and that he, Forbes, was the better known of the two. Both James and David Kinley, who would replace James as president of the university of Illinois in 1920, had met biological scientists abroad before and after the turn of the century, who upon learning that they were from Illinois would immediately say, "of course that is where Professor Forbes is."

At President James's reception in honor of Forbes's resignation that June, Burrill said:

> Dean Forbes has filled the prominent position with great acceptability everywhere and to everybody . . . he has been faithful in duty, wise in counsel, courageous in action, admirable in association. . . . The College of Science specifically suffers the most direct and greatest loss, but may also most directly share in the gain since scientific accomplishment is preeminently its goal.

When word reached the public of Forbes's intention to retire his deanship, some believed that he would no longer be with the university, a supposition that was not refuted until June 8, 1905, the day after he received an honorary Doctor of Laws at university commencement exercises. Characteristically, no sooner had Forbes retired as dean than he assumed the position of chairman of the Committee on Educational Policy of the University Senate, a position he then held for sixteen years. He later received this accolade: "To his ability, poise, sagacity, and progressive thought may be attributed much of the educational advancement which the University has made."[6]

After his retirement as dean Forbes had no let-up either in his entomological inspection, research, and education work. His activities once again caught the attention of L. O. Howard in Washington: "in no state has inspection and remedial work been carried out more thoroughly than in Illinois." For example, San José scale, after more than seven years of diligent orchard inspections, still had not turned up in almost half of the counties in Illinois. Forbes's success in his endeavors led predictably to significant expansion of duties by his office under new state laws in 1907 and 1909. These laws clarified and strengthened earlier legislation by extension to injurious insects and fungi of livestock, truck farms, vegetable gardens, shade trees, ornamental plants, products of mills and warehouse stores, and other situations concerning public health and hygiene (e.g., flies carrying diseases). State appropriations increased accordingly, providing him with more than $20,000 annually. By 1910 his entomological office had thirty people: himself, ten entomologist assistants, a draftsman, a chief inspector, four subinspectors, a foreman for insecticide operations, and twelve laborers—not to mention a legion of involved farmers and horticulturists as allies.

Almost simultaneous with this expanded work came the chinch bug explosion of 1910–14. It caused an estimated $13 million loss in yields of corn, wheat, and oats from twenty-three counties in southwestern and southwest central Illinois. But it could have been far worse had it not been for Forbes's and his field crews' leadership, the cooperation of hundreds of farmers, businessmen, and newspapers, and—luckily—welcome wet weather during bug breeding periods in 1912 and 1915.

Forbes believed that with a vigorous plan of attack using proven methods, and with special state appropriations, that the chinch bug could be held in check. The key was to attack the first generation of young wingless bugs marching on foot out of infested wheat and oat fields at midsummer harvest time in their search for other food crops—especially corn. Coal tar and creosote barriers, supplemented by posthole traps around the infested fields would do that, as Forbes's field crew had shown: they easily captured bushels of bugs in this fashion, emphasizing that a bushel held some 8.5 million young bugs. Any bugs escaping on foot or later flying into corn could be controlled by spraying tobacco extract or soap solutions.

Use of these recommended methods saved thousands of acres of corn from injury and showed that even better gains would have resulted given more extensive cooperation. For example, close study during the final campaign of 1914 in thirteen highly participatory counties of west-central Illinois showed significant maintenance of corn yields behind 1,500 miles of protective barriers and traps against costs of materials and labor of less than $50 per mile.

This was, Forbes trumpeted, the first time in any large Illinois district that farmers had "really undertaken to defend their crops against the chinch bug" instead of stoically waiting out depredations. They were members of the biological association of the cornfield and had to rationally participate. Cooperation in the common interest had worked and once again demonstrated Forbes's powerful influence on agricultural matters in Illinois.[7]

Forbes's biological research, scholarly writing, and steadfast service to society led to his election as president of both the newly formed (1907) Illinois Academy of Sciences in 1909–10 and—for the second time—of the American Association of Economic Entomologists in 1908–9. At age 64 in late 1908, he prepared his presidential address for entomological colleagues meeting in Baltimore with the AAAS. Almost half of the address would consider an ecological viewpoint and methods for improving their work. He had been practicing ecology during its formative years as a science since his formal recognition of it in 1894. So seriously did Forbes take this matter, that he would argue that "ecology evidently lies at the very center of biology." In truth he felt that few colleagues would be overly surprised at what he would say. For years, of course, Forbes had consistently pushed for ecological approaches with them, but he wanted to get his remarks right, in emphasis and tone, and aired in a societal forum.

Forbes began that December morning by noting a gratifying increase since 1893 in the number of well-trained men working in economic entomology, using improved methods of scientific investigation and publication. He then went on to forecast "the probable next steps in the development of our work." Reminding his colleagues that economic entomology and agriculture can both be unpredictable, and that as scientists they observed, generalized, experimented, and verified in order to predict, he emphasized that long-term studies and repeated experiments supported by statistics would markedly improve their predictions.

Furthermore, he had looked in recent years "with eager interest to the new and still developing *methods of ecology* for helping solve our larger and more difficult problems." He now believed that "economic entomology is, in fact, a special division of ecology." As such, this special field dealt with the interactions between insects and humans, and the effect of these interactions on human health and welfare. Hence, whether they knew it or not, they had been in the forefront as bona fide working ecologists. Yet he believed they had been too strictly practical and needed an even wider ecological view of their work. Here they could learn from the current work and methods of "a group of active young ecologists, who unfettered by any responsibility for an economic result," were freer to inquire into the history and development of the distribution and abundance of groups of species as a result of their dynamic interactions and adaptations.

This knowledge should, in turn, help economic entomologists treat their problems in a broader context. For example, information gained about the probable relations between insects and "the organic world at large before civilized man appeared" could put into perspective the general impact of human kind, and provide the costs and benefits of modifying or disrupting the primitive natural order which humans "can never really replace."

Forbes then illustrated an ecological view as applied to the biological association (or biocoenose) of a cornfield. The dominant associates are the grasses, including corn itself, the corn field ant, corn root aphid, corn rootworm, white grubs, ear worm, and finally the horse and humans. Then there are the occasional (sometimes voracious) or less conspicuous associates: other insects, birds, grasshoppers, mice, and ground squirrels. The corn plant, humans, and the horse essentially oppose all the others. The humans strive to destroy the other animal dominants as best they can. But the mere fact of the cornfield's existence, with the incessant aid of humans, allows the ant, aphid, and root worm to thrive there more than they would on other plants elsewhere. Finally, the seasonal succession of plants in the cornfield make it possible for the aphid to continue there: the aphid feeds on early spring weeds before the corn is sown. The cornfield was an ecological system and should be studied in that fashion.

Forbes considered the corn plant as "permanently fixed in a state of infantile helplessness," made so by selection of characters important to human needs. He suggested instead that someone make the experiment of growing corn from seed taken from the best stalks in a field either overrun by insects or stressed by drought. In short, they should select those plants best withstanding *unfavorable* conditions, and perhaps put the corn plant on the road toward greater hardiness.

Lest his colleagues become overly enamored with the pure entomological part of their duties, Forbes warned that they could only "rest satisfied" when "the conditions of life for our people . . . have actually been improved." And that would happen only when "we, or someone else for us, can hitch fact to practice," a result they could obtain by helping each other. In Illinois he relied on an advisory committee—with men from the experiment station, State Horticultural Society, and Farmer's Institute—formed to consult and consider his plans after full discussion. The committee either tested entomological results itself or influenced others to try them, often by way of demonstration stations.

Finishing up, Forbes briefly addressed the need for both education and enlightened law in protecting against the spread of insect and fungus pests for the nursery trade, fruit grower, and farmer. "Dangerously infested property is a nuisance, and in my judgement should be universally so treated." He urged his colleagues to help strengthen laws in their several states. They had become in recent years, "a guardian of the public health as well as a protector of its property. . . . By reason of our past achievements, the country is coming to expect more and more of us, and is yearly more willing to enlarge our opportunities and support our undertakings."

Forbes's address was well-received by the more than one hundred members present—the largest turnout the association had ever had. "No one who heard or read his address will ever forget it," wrote Leland Howard. "It made a very deep impression which has resulted in no end of good to all of us in our work and in our view of things." Beyond question, Howard thought, Forbes had established his leadership because of "his broad experience in biology [and] his broad outlook on nature."[8]

During the summer of 1909 Forbes prepared for his resignation from the zoology department and his assumption as head of the fully fledged entomology department, a move that was accomplished by September 1. He was especially anxious to get new upper-level entomology courses rolling, ones that included the ecological content he earnestly felt his students needed. Zoological ecology—a lecture, laboratory, and field course—would be taught with Folsom and Amos W. Peters; faunistic zoology—a graduate lecture, field, and conference course covering the taxonomy, distribution, and ecology of fish, birds, and mammals—would be taught with Folsom and Frank Smith.

Forbes and Folsom alone continued to make up the faculty of entomology through the two academic years from 1909 to 1911, the younger man again handling the five lower-division courses, and both handling upper-division undergraduate and graduate courses. The latter included faunistic and ecological entomology covering taxonomy, distribution, and ecology of insects, and an independent research course in ecological or economic entomology. These graduate courses, and a course in advanced economic entomology, were enhanced by operations, equipment, and collections of both the state entomologist's office and the expansive facilities of the State Laboratory of Natural History, especially during the years before 1916. In 1910 entomology matriculated eighty-five students, thirteen of whom were graduate students.

Courses in entomology nearly doubled by 1914 with the addition to the faculty of Dr. Alexander D. MacGillivray, a systematic entomologist (sawflies and scale insects) from Cornell, and Robert D. Glasgow, an economic and medical entomologist, who had completed his doctoral degree with Forbes in

1913. At age 70 Forbes had by then ceased teaching undergraduate courses, except for his sponsoring independent study in economic entomology for advanced undergraduates until his seventy-second year, the year after he said farewell to graduate courses, too.

While Forbes was department head from 1909 to 1921, his objectives were to prepare students for: professional work in economic entomology, entomological teaching and research, teaching of natural science and entomology in the public schools, and a minor in natural sciences including entomology. This quadruple set of mission tasks was reflected in an increase in courses averaging eighteen during 1915–21, and the participation of two new instructors, Drs. Edna Mosher (Ph.D. under MacGillivray, 1915) and Charles P. Alexander, entomologist, State Natural History Survey, and a half-dozen graduate assistants.

Forbes was 63 when a graduate degree program was organized and fully formed at Illinois in 1907, and no Ph.D. degrees were awarded in zoology or entomology before 1909. Forbes himself sponsored only one master's degree candidate in zoology and sponsored only three master's and two doctoral degrees in entomology (see appendix 3). During his tenure as head of entomology, twenty students received master's degrees (1908–19) and nine received doctor's degrees (1912–20).[9]

Just before and again soon after his move to entomology, Forbes continued to gather honors. In 1908 he was elected a fellow of the AAAS, and in June 1910 to Phi Beta Kappa. Then for 1912–13 he was elected president of the six-year-old Entomological Society of America, automatically also serving as a delegate to the Second International Congress of Entomology during the summer of 1912 in Oxford, England.

Stephen and Clara sailed from New York in late July on board the S.S. *Kroonland*, arriving in the English Channel on the twenty-ninth. From the time of their arrival through August, they traveled through London, Dover, Canterbury, and Oxford. His visit even included a meeting with Rothchilds. Forbes wrote his son Ernest that "the Congress was about as satisfactory as such things usually are." More Americans were in attendance "than any other group except British." Forbes had presented a paper that was warmly received, concerning possible transmission of contagious disease by black flies. Yet Forbes felt it had been "an imperfect discussion of a problem yet unresolved."

At issue was Dr. Louis W. Sambon's theory of transmission of pellagra via microorganisms by biting flies of the genus *Simulium*, better known as black and buffalo flies, as laid out by him in the *Journal of Tropical Medicine and Hygiene* in 1910. Forbes's involvement came with his appointment by the governor of Illinois in 1910 to a state commission investigating the outbreak of pellagra in psychiatric asylums and other state institutions. In fact, 63 percent of the known 408 Illinois cases during 1909–11 centered in the Peoria Hospital for the Insane, and 30 percent more were from three other urban asylums—all four hospitals drawing their patients primarily from the Chicago area.

From his work in Italy, Sambon noted peaks of new pellagra cases in spring and fall coincident with two generations of flies. The disease was prevalent in rural mountain valleys, very different from Illinois, where 93 percent of cases were from in or near cities.

Forbes considered *Simulium* to be ubiquitous in Illinois reaching almost everywhere beyond their breeding waters, but falling off rapidly in abundance by July. Yet his Illinois data showed no correlation between the seasonal abundance of the insects and the number of new cases of the disease. Consequently, he felt obliged "to take a critical attitude toward the *Simulium* theory of this disease," although admitting that his limited Illinois data were "not, by themselves, conclusive either for or against" Sambon's reasoning. Forbes's stance was later supported when pellagra, marked by skin eruptions, digestive upsets, and nervous disturbances, was shown between 1934 and 1938 to be caused by chronic niacin deficiency and not by microorganisms carried by an insect.[10]

Leaving the south of England, the Forbeses trained north to York, Newcastle, and on into his ancestral Scotland, spending a week in Edinburgh "having the time of our lives." Soon after they sailed the Channel to Le Havre, then went by train to Paris. During September they toured Germany, visiting Berlin, Potsdam, and Dresden, where Forbes consulted with German biologists working on the Elbe River. They sailed for home on the S.S. *Finland* from Hamburg, landing in New York early in October 1912, tired but happy.[11]

By 1912 the U.S. Bureau of Entomology had begun a marked expansion in the amount and variety of their fundamental work in economic entomology. This in turn increased research opportunities for the state agricultural experiment stations. Besides, in 1913 the bureau now operated thirty-five field laboratories in various states, up from only one field lab sixteen years earlier. Within five years the bureau's annual budget would surpass one million dollars, up from less than one-tenth that in 1906. Moreover, passage of the federal Smith-Lever Extension Act in 1914 relieved the experiment stations from an excess of routine work, so that researchers could devote themselves to the entomological investigations they were uniquely qualified to do. For Forbes, as he wound down his last few years as state entomologist, these included studies of the codling moth, Hessian fly, May beetles with their white grubs, and shade tree insects, the details of which he added to his storehouse of applied knowledge on the interactions between insects and humankind.

As immediate past-president of the Entomological Society of America, Forbes was invited in 1914 to present the annual address to the society in Philadelphia on December 30 at AAAS meetings in the Walton Hotel. The subject was the ecological foundations of applied entomology, and he made recommendations about how entomologists and ecologists might work together for each others' and society's benefit.

"The very substance of applied entomology [is] ecological through and through," Forbes asserted, with the primary area of interest being the dynamic one, or what the organisms *do*. This was the causative factor for the static (or distributional) and successional relations of organisms, including the human species. Specifically, applied or economic entomology "is that part of ecology over which the ecology of man and that of insects is coincident."

Indeed, Forbes took issue with those people—ecologists or otherwise—who wrongfully considered human beings "outside the natural system," who considered primitive nature as an "earthly paradise," and civilized humans "a kind of fiend," whose entry into the ancient world "had introduced . . . the germs of that fatal and frightfully contagious disease known as civilization." Forbes believed that an ecological view "must include the twentieth century man as its *dominant* species—dominant not . . . as simply the most abundant . . . but dominant in the sense of dynamic ecology, as the most influential, the controlling or dominant member of his associate group."

Practically all the useful things that applied entomologists needed to know concerned the ecology or physiology of injurious insects. Even the human factors in economic problems, he thought, were really ecological in character. They concerned reactions and interactions among humans themselves in response to insect depredations, responses that were often not in the general community interest, because "mankind has not put its heart effectively into the business." This was as true for chinch bug control as it was for malarial disease and mosquitoes, where for fifteen years after the important scientific discoveries were made it was easier to control the malarial parasite and the mosquito than it was to persuade potential human victims to make proper use of these control measures.

For their purposes entomologists had, without hesitation, recently carried out important ecological work in the physiology, reproduction, and life histories of insects. Yet they were also prepared to learn from the pure ecologists about the physical environment of insects, the other organisms with which insects lived, and how particular habitats and their faunas had changed through time. In this fashion economic entomologists and pure ecologists could address both the present and the past, thus preparing for the solution of future problems in their common interest.

In conclusion Forbes recommended that university students of economic entomology take courses in general ecology and that students of ecology take courses in applied biology. Both groups might then achieve a much-needed broader outlook and a new point of view.

On the same day Forbes gave his address, two dozen professed ecologists met informally that evening in the hotel lobby to consider the formation of an American ecological society. Conceived as a way to bring animal and plant ecologists together for discussion and fieldwork, the idea rose from correspondence between zoology professor and physician Robert H. Wolcott of the University of Nebraska and Victor E. Shelford, newly hired assistant professor of zoology and head of the new research facility of the State Laboratory of Natural History at Illinois. Substantial agreement that night for such a society led to appointment of an organizing committee and a call for a conference of ecologists at the December 1915 AAAS meetings in Columbus, Ohio. There, with positive votes from fifty in attendance, and from a like number of letters, a society was formed, a concise constitution adopted, and a slate of officers elected. The first officers of the Ecological Society of America were: Victor E. Shelford, president; zoology professor William M. Wheeler of Harvard University, vice president; and plant ecologist Forrest Shreve, Desert Laboratory, Arizona, secretary-treasurer.

Early in 1916 Forbes made known to Shreve his intention to organize his Illinois colleagues as a "section" of the society. In December at Columbia University and Barnard College in New York City, the society met in conjunction with the American Association for the Advancement of Science. The society listed 284 charter members and 23 newly elected members. Thirty-two of the total 307 were from Illinois, the highest ranking state, and 39 identified themselves as entomologists, the fourth-ranked class behind plant ecologists, animal ecologists, and foresters. Forbes presented a shortened, fifteen-minute version of his 1914 address to the Entomological Society to his ecological colleagues there in New York and then proposed a committee to review and report on the salient connections between ecology and economic entomology, and how they might work cooperatively in their common interest.

Forbes stepped down from his state entomologist's position in 1917 after thirty-five years in office, and with little fanfare. He would retire from the university faculty and as head of entomology four years later at 77.[12]

CHAPTER 7

Aquatic Systems, 1880–93

F orbes's first paper of 1880 on the foods of fishes, the reader may recall, had the objective of examining possible correlations between the anatomy of a fish's mouth, throat, and gills and the kinds of food it ate. But the introduction to that paper also dealt with how animal foods served as an important focal point in the ways animals interact both with each other and with the activities of humans. Forbes called this "economical biology" and discussed it fully in his theoretical "Interactions" paper (1880). Also in that introduction, Forbes called the theater in which the play of interactions took place a *microcosm*, or small world, representing a much larger natural system, but small enough Forbes wrote, "as to bring it easily within the mental grasp." Forbes cited as examples a lake, river, or stream. The animals and plants there formed an "organic unit," displayed interactions, and were dependent on all the "elemental forces" (or conditions of existence). In this fashion Forbes anticipated the ecosystem concept fifty-five years before the term was coined by the British plant ecologist Arthur Tansley.

Reputedly, Forbes found the German zoologist Karl Möbius's 1877 concept of a biocoenosis (community of an oyster bank) useful in his own description of the microcosm, with both sharing interacting elements and tendencies toward equilibrium. Forbes never denied this influence, according to animal ecologist Victor Shelford, a colleague of Forbes at Illinois from 1914 to 1930. Forbes appears innately holistic in his thinking, and he surely was strengthened as well by his eclectic reading habit since childhood and his close relationships with such like-thinking colleagues as Thomas Burrill, Cyrus Thomas, David Starr Jordan, Charles Kofoid, and Leland Howard. Consequently, Forbes microcosm concept was also instrumental in the development of the biotic community concept.

During 1880–82 Forbes and his assistants made trips to lakes in northern Illinois, southern Wisconsin, to Lake Michigan, and the lower Illinois River. There they collected with dredge, beam trawl, and surface plankton nets for fish and aquatic invertebrates, and took soundings, water temperature, and bottom samples for further analysis. Their purpose was to gain information on species' foods, their conditions, and the system of relations by which the groups of organisms are "held together." Forbes frequently said that this information was of potential value to people of both pure scientific and practical interests, for example, to both research biologists and fish culturists. Forbes, however, did not explore this microcosm concept again in published form during the next half-dozen years.

In mid-August 1884 Forbes received a request from Spencer Baird, U.S. Commissioner of Fish and Fisheries, to investigate a massive one-month long mortality of fish in Lake Mendota, Madison, Wisconsin, home of the University of Wisconsin. Here, Philo Dunning, head of the Wisconsin State Fish Commission, provided Forbes with a workroom at waters' edge. Here too, Professor Edward A. Birge, a colleague of Forbes, taught zoology and studied cladoceran crustaceans. Lake Mendota, the fourth largest of Wisconsin lakes, measures 9.2 km (5.7 mi.) long (east to west) by about 5 km (3 mi.) wide, and is about 21 m (69 ft.) in its deepest trough. Some 90 percent of the 300 tons of dead fish carried away by Madison city authorities that summer were yellow perch *(Perca flavescens)*, with lake herring or cisco *(Coregonus artedi)* next abundant, followed by small numbers of pike perch, white bass, and sunfish.

Forbes stayed at Lake Mendota a week, taking dredge, sieve, and surface net hauls for the fauna with aid of a steam launch. He expanded these collections with his assistant H. W. Garman the following summer. He also collected dead and dying fish, made fish autopsies, sampled stomach contents, and prepared bacteriological slides and tissue samples excised from seven organs for microscopic analysis. When analyzed, these bacteriological and histological samples from perch and herring gave no clue of disease.

A large majority of dead perch were mature fish. They were healthy looking, plump, and displayed bright normal color. Forbes could not find any trace of fungus, parasites, abnormal blood, or anything unusual about the gut contents that were usually well-filled, recently eaten, and almost all consisting of long red *Chironomus* midge larvae (20–30 mm [.8–1.2 in.]). These larvae appeared only in deep-water dredge samples from mud bottoms. For comparison, Forbes sieved healthy perch from near shore. Gut contents from these fish contained chiefly shorter (10 mm or so) white *Chironomus* larvae of a different species, in addition to small crustaceans, mayfly nymphs, a leech, caseworms, a mollusk, and some algae—an assemblage similar to that found in Forbes's shallow water dredge samples.

In conversations with fishermen and others Forbes learned that smaller numbers of herring "died every summer." Most knew, too, by catching them there, that mature herring and perch commonly fed on the bottom. He also pondered reports of a flooding rain shortly before the fish kill, thinking perhaps that large amounts of unsavory organic material may have swept into the water from the 80 acres of marsh circling the lake. Kills of fish under similar conditions were common in Illinois rivers during summer. Whatever the cause, Forbes wrote, both perch and herring perishing in Mendota "were ranging in the deeper water, [with] their last meal of insect larvae found only in the mud of deeper parts of the lake."

No one, including Forbes and Birge, knew what caused this fish kill in 1884. Not until 1898 did Birge report on the summer thermal stratification of Lake Mendota, and then in 1910 expound on factors controlling the oxygen content of deep lake water in the summer. Below the thermocline, or the sharp line low in the lake where temperature dropped more rapidly with depth, oxygen disappeared quickly and often completely in summer, used up by the oxidation of large amounts of decomposing organic matter raining down to the bottom. And in the bottom mud that Forbes described as "stinking" when brought up to the deck of his steam launch, lived larval *Chironomus*, true worms, and other invertebrate animals which Birge's colleague, Chauncey Juday, showed in 1908 (and in 1921 by analogy) could either remain active in summer months during a lack of oxygen or else become dormant. So here was the food attracting the deep-ranging fish and reachable by them during the summer only at seemingly great risk.

Despite his inconclusive findings, this episode shows Forbes's willingness to materially assist federal efforts in problems concerning public waters. This skill would continue to bring him professional opportunities and financial support, while simultaneously increasing his understanding of aquatic systems.[1]

For the next few years Forbes's zoological work at the state laboratory picked up on "the aquatic animal life of the state," which he studied "systematically both in detail and as a whole." The work included the usual distribution, habits, and food of the fauna, as well as their relations to water quality and human food, and "to each other where they are thrown together, as in the same lake or stream." Moreover, his work came into closer, more mutually helpful contact with that of the Illinois State Fish Commission. He was increasingly convinced of his earlier comment that animals and plants in a region needed to be treated as "an organic unit . . . in action," and with "principal attention to the laws of its activity." Ultimately this was needed whether the purpose of the work was economic, educational, or scientific. In any event Forbes was gratified to receive scientific acclaim for his work with his receipt in 1886 of a first-class medal from the Société d'Acclimatation de France for his interactions paper and his series of papers on the foods of birds, fish, and insects.

In the spring of 1886 Forbes was invited to address the Peoria Scientific Association that coming winter. He had led occasional natural history excursions for this group, a break that he readily enjoyed from his routine. The association had a museum and lecture room in the spacious Peoria courthouse, where all regular meetings were free and open to the public. Membership was five dollars annually. The naturalist J. F. Kennedy was curator and Forbes a "corresponding member." Forbes's subject was "at his own choosing," prompting him to present the concept he had been mulling over for six years: the biological system of freshwaters. (The topic he chose raised such keen interest in Peoria that he repeated it in June 1887 at the commencement exercises of Indiana University by invitation of his friend President David Starr Jordan.) Forbes titled his talk "The Lake as a Microcosm."

Forbes spoke eloquently the night of February 25, 1887, in Peoria. He said that a lake was "a little world within itself," with a "dynamic equilibrium of organic life and activity." Whatever affects any species in a lake "must have its influence of some sort upon the whole assemblage." Forbes listed the physical characteristics and compared the fauna of different sized lakes he had studied from Illinois and Wisconsin with some from Europe.

Then in his inimitable way, Forbes said he would provide the "pigment" for his audience to paint their own pictures of lakes and their interacting living creatures. Characters in their paintings would include animals that swam, crept, crawled, burrowed, and climbed through the water, "in and on the bottom, and among the feathery water-plants with which large areas of these lakes are filled."

Next Forbes eloquently described the food web interactions that shaped the species and the assemblage or communities of plants, invertebrate animals of the plankton and benthos (bottom), and the fishes. The foods included those he had exhaustively studied in his fish work: the Entomostraca, midge larvae, water bugs, and other aquatic insects; shrimps and crawfish; mollusks; and the fish themselves. Finally, he had found not only young fish, but insects, larger crustaceans, and even the carnivorous bladderwort plant *(Utricularia vulgaris)*, all feeding on Entomostraca, those "truly exquisite" tiny crustaceans such as copepods, ostracods, and cladocerans, showing the dependence of many species on similar foods.

He closed by theorizing as he did in his interactions paper in 1880: natural order had evolved in this microcosm through a dynamic balance of species abundances within limits of the food supply, an order maintained by natural selection to the mutual benefit of a mere assemblage of plants and animals and now "organized as a stable and prosperous community."

Forbes concept of a lake as a microcosm, and as "remarkably isolated," was a simplification and would change as ecologists gained a better understanding of watershed, atmospheric, and human inputs to a lake. His holistic view of a lake as an "organic system" was a classic contribution to limnology and ecology. It acted as a guiding principle for most early limnologists, including Birge, F. A. Forel (the "father" of limnology), and A. Thienemann, and for many others into the second half of the twentieth century.

Forbes clearly viewed a biological system as action-oriented, and he recommended attention to learning "the laws of its activity." A careful reading of this 1887 paper shows that he perceived a reciprocal shaping relationship between the parts (species, in this instance) and the whole (the microcosm), and it was through the interactions of the species that their roles were more clearly defined. Thus Forbes's contribution has remarkably wide utility for limnologists, ecologists, and others with interests in different parts and workings of biological systems. The microcosm paper would have a great influence on biological thought both in the United States and abroad.[2]

During the summer of 1888 Forbes collected fish from the Illinois and Mississippi Rivers to conclude his long-running food studies. Meanwhile, Garman studied the Mississippi River bottoms near Quincy, Illinois, where Garman lived, in the Illinois State Fish Commission wharfboat and worked from his own smaller boats with several helpers. Garman detailed the animal assemblages of benthic habitats where young fish were collected by state workers for redistribution around Illinois. He paid close attention to faunal differences between the open moving river and quieter backwater sloughs, ponds, and "lakes" of the flood plain, where with varying amounts of isolation and drying up during summer, the fish and benthic fauna were often cut off. Several Mississippi valley states and the U.S. Bureau of Fisheries accordingly had carried out "fish rescue" operations beginning in the 1870s. Illinois, however, began in 1890, and ceased in 1910, some 10–15 years earlier than other agencies in part because of Forbes's opinion on the matter. He had suggested in his microcosm paper that the mortality of backwater fish under naturally fluctuating water levels might be beneficial over time as a mechanism holding fish species like minnows, bull pouts, pickerel, sunfish, black bass, and perch within reasonable limits of a uniform abundance year after year.

Even more important for cessation of fish rescue operations, at least on the

Illinois River, was the mounting restriction of public access to the floodplain of that river and its enclosure by private interests beginning in the late 1880s. Fishing and hunting clubs, large fishing companies, and agricultural interests bought land in the floodplain as it was increasingly leveed and drained, especially between 1900 and 1910. In parallel with Forbes's development of his aquatic research program, the state of Illinois attempted to balance drainage and conservation interests as small fisherman, hunters, and other residents of the floodplain fought for their historical common property resources in fish, game, timber, and nuts.

After a dozen years' work on the food of fishes from rivers, lakes, ponds, streams, and temporary pools, in 1888 Forbes summed up his studies. They were based on more than 1,200 specimens (914 adults, 307 young) of 87 species of fish, giving the relative proportions of the different types of food eaten as a measure of their importance. For adult fish, insects (40 percent of foods) remained first even with the large increase in fish sample size since 1877, followed by fish, crustaceans, mollusks, and plants. *Chironomus* and Entomostraca together remained first in importance for all young fish, making up 75 percent of their diets. Despite the more obvious difference in foods between fish with quite different alimentary structures as described earlier, Forbes noted dietary differences as well between some fish species with similar structures and close taxonomic position. Sometimes their food searching habits were different; if not, their preferences were obscure. Forbes called these psychological preferences and presumed they were heritable and responded to natural selection.

The work was tedious but rewarding, the data voluminous but interesting to sort out. Forbes had genuine fun with it, and sometimes there were surprises: catfish with their sharp spines were eaten more frequently than expected; although the larvae of surface water beetles were eaten, the whirling, darting swarms of adults commonly seen were not found in fish stomachs; plants made up less than 10 percent of foods; and true worms made up only a trace of fish food. The latter point is perhaps not surprising to readers who fish—which Forbes did not—and know that it is the wiggle of the plump terrestrial worm on the hook that gets the strike. (Worms this size do not occur naturally in fresh water, as Forbes knew, which may indicate a psychological preference of fish for plump wiggly things.)

Earlier in the spring of 1888 Forbes ventured to lay out his ideas on fish foods, the ecology of fishes, and the relevance of his work to members of the American Fisheries Society at their annual meeting in Detroit. Not pretending to be a fish-culturist, which many in the society were, Forbes said that he could, however, assist in their understanding "the system of aquatic life into the midst of which we thrust our little fishes." Mistakes were commonly made in not knowing the limits of natural waters for supporting the coveted fishes that were stocked. Likewise, foreign fish species had been introduced without knowledge of their food requirements, or their effect on native species. Furthermore, "We cannot plow and till our lakes and rivers as the farmer does the prairie sod, ruthlessly exterminating all of the native forms of life in order to substitute other sorts more useful to him. . . . If we cannot get rid of the natural order, we certainly need to understand it."

Forbes said that he could now provide the food habits of many fish species at all ages and under a variety of environmental conditions. This information could serve fish culture, introductions, stocking, and fisheries management; without it money would be wasted and reliance would continue to be placed on "guess-work and empiricism." Society members were vividly impressed with Forbes's report, the first comprehensive study of its kind and scope ever performed with fish foods.[3]

By the fall of 1888 Forbes's habit of adding on more and more duties and activities during the past several years began taking a toll on his energy. He now also lost two of his most experienced assistants: Garman to the Kentucky Agricultural Experiment Station, and C. M. Weed to the Ohio Station because of "inadequate provision for their salaries." Consequently, his general zoological work, including aquatic biology, slowed down.

For relief he decided to take a relaxing, personal, weeklong zoological excursion to northern Michigan in the summer of 1889, one urged and supported by the U.S. Fish Commission. His prey were entomostracan crustaceans—copepods and cladocerans, their kinds and abundance, and their relationships with the fauna of European lakes then under intense study. His collection points were l'Anse, Marquette, and Whitefish Point, on the south shore of Lake Superior, and delightful Lake Michigamme, west of Marquette in the Menominee River watershed emptying into Lake Michigan. Previous work in Superior consisted of a single report by S. I. Smith of Yale in 1874, which mentioned four species of these tiny elegant crustaceans. Forbes used a small net from piers and breakwaters, and a skiff in Lake Michigamme, collecting to 4–12 m (13–39 ft.) depths mostly during daylight hours.

Forbes came up with two dozen species, including three new to science: two copepods, *Cyclops gyrinus* and *C. edax* (now *Mesocyclops edax* [*Forbes*]), and a new cladoceran, *Chydorus rugulosus*. Close examination of six other species of cladocerans showed these animals were essentially identical with descriptions of similar species from European lakes, and indeed are now considered distributed widely in North America, Europe, and Asia. Two calanoid copepods, *Di-*

aptomus sicilis and *Epischura lacustris*, both described as new species by Forbes in 1882 from Lake Michigan, now appeared in Superior and the latter species in Lake Michigamme. Both species were closely related to crustaceans from European lakes. Forbes's work, and that of E. A. Birge in 1892, were early contributions to the growing realization of a wide intercontinental distribution of many cladoceran and some copepod species, as well as other North American species and genera from these crustacean groups that clearly had close relatives in the Old World. The common faunal origins expressed here are based on presumed closer ancient connections between freshwaters of the upper northern hemisphere. These connections were later severed by glaciation, followed by morphological differences arising in some species during the isolation of North American populations from their European and Asian counterparts.

Through colleagues in the U.S. Fish Commission and David Starr Jordan, Forbes stayed abreast of aquatic work around the country. While he hauled his collecting net for elegant Entomostraca in Lake Superior, Jordan was busy in Yellowstone National Park identifying which lakes in that Rocky Mountain country lacked fish and why. In 1880 the commission had begun stocking some lakes with trout but without knowledge of the kinds and amounts of fish food available. It turned out that lakes were fishless because of physical barriers to fish passage from below, such as falls or steep topography alone. Jordan's quick checks showed diverse invertebrate animals present, but a more experienced researcher was needed to complete the work. As Richard Rathbun, assistant for scientific inquiry for the commission would say, freshwater invertebrate work "has been entirely neglected by the Fish Commission," and "there is no one else in the country who has had [more] experience in the freshwater invertebrates" than Forbes. So it was Forbes to whom Col. Marshall McDonald, new U.S. Commissioner for Fish and Fisheries, now turned for help. Forbes would inform Rathbun that he considered himself "a steadfast friend" of the Civil War veteran McDonald and reaffirmed his willingness to help in commission investigations.

During the spring of 1890 Forbes made arrangements for a summer trip west, obtaining cost estimates and suggestions for equipment and logistics from Jordan. He arranged with a guide to visit "all parts of the Park." He then wrote the governor of Illinois, Joseph Fifer, to explain his impending two-month summer absence from campus. His physicians had warned him that he must "either take the longest possible vacation or prepare for a complete change in the character of my business." Forbes's solution would be more personally palatable than what the doctors recommended: he would take his chosen business completely outdoors into Wyoming's bracing mountain air. Forbes's field associate would be Professor Edwin Linton of Washington and

Jefferson College, Pennsylvania, an expert in the parasites of fish. The rest of his party consisted of a guide, a teamster, two packers, and a cook. Their outfit included camping gear and food, pack mules, saddle horses, a canvas boat, collecting gear, microscopes, and preserving chemicals and glassware.

Departing Champaign-Urbana on July 11, Forbes reached Mammoth Hot Springs in Yellowstone Park on July 17 and set up camp that evening on Swan Lake plateau in the northwest corner of the park. Over the next six weeks his party would trace a winding loop on the central-western plateau, visiting forty-three collection sites to an altitude as high as 2,500 m (8,200 ft.). He would leave the northeastern part of the park for the following year. Forbes's understanding with McDonald emphasized the fish food angle of his study, and it was agreed that a general zoological survey of park waters would also be performed. The location of the park smack on the continental divide lent further interest to the work and promised valuable material for faunal comparisons with lower altitude waters of the region. Most important, it allowed Forbes to test his microcosm concept and determine the roles of animals in the "biological system" of mountain lakes.

The continental divide here runs northwest and southeast through the southwestern quarter of the park, less than 2–8 km (1–5 mi.) from the western and southern borders of Yellowstone Lake, the largest lake at the summit of the Rocky Mountains. Forbes's collections were made from the Gardiner, Madison, and Yellowstone River systems on the Atlantic slope of the divide, which drain the northern and eastern sections of the park, and from the Snake River system on the Pacific slope, which drain the southwestern section. He collected with dredges, sieves, fish trammel net, plankton nets, and small hand nets from shallow water to 60 m (197 ft.) depths.

Forbes's report brims over with detail of the fauna, precluding easy summary. What follows here relates primarily to five interesting mountain lakes all near 2,500 m in altitude, either with or without fishes, and with details on their plankton and some additional crustaceans. It should be understood that Forbes's work overall dealt not only with lakes, but also with ponds, creeks, and rivers.

Leaving camp by saddle horse on July 18, his party rode south 40 km (25 mi.) through Norris Basin, down the Gibbon River, and the next day reached the Firehole River. Here the teamster and saddle horses were exchanged for mules and a wagon for the trip across the divide through Norris Pass, and on to the north end of fishless Shoshone Lake (11 km [7 mi.] long) in the Snake River drainage.[4]

"It seems to me," Forbes wrote his sister Nettie, "this must always count as one of the most charming recollections of my life," as they rode the Shoshone

Trail, came out of the trees and beheld the vista of the clear blue lake before them. The lake lay in a hollow, he wrote Clara, "surrounded by pine-covered mountains . . . rimmed by black volcanic sand, and a few miles away the dark, bare, rocky peaks of the Red Mountains loom upward . . . half-covered by snow," with Mount Sheridan commanding that southeastern rampart.

They hauled a net from the canvas boat for plankton all day for three days and dredged from 2–12 m (7–39 ft.); they took shallow water and shore collections and sounded the north bay. Forbes reflected that no one had done this much collecting here before. Shoshone Lake as well as the other four lakes all yielded diverse Entomostraca, *Chironomus* larvae, leeches, mayflies, and other aquatic insects, and mollusks. But there were important differences. At fishless Shoshone, two species of amphipod crustaceans, now known as *Gammarus lacustris* and *Hyalella azteca*, were everywhere, the former species even on the surface at night. And then there was the "gigantic" (2.6–3.6 mm [.1–.14 in.]), scarlet, predaceous copepod that Forbes described as a new species, *Diaptomus shoshone*. It occurred in large numbers in the plankton and had copepod remains and pine pollen grains in its gut. Two other much smaller herbivorous diaptomids also turned up in the plankton, *D. sicilis* (1.1–1.5 mm [.04–.06 in.]) and another new species, *D. lintoni* (1.5–2.0 mm [.06–.08 in.]), as well as a mixture of other copepods and cladocerans. Forbes said that "as far as crustacea were concerned, the lake was in practical possession of amphipods and *Diaptomus shoshone.*"

On July 24 the party left Shoshone for fishless Lewis Lake, 19 km (12 mi.) south, and then worked this 5 × 4 km (3 × 2.5 mi.) lake for two days as best they could in stormy weather. The two amphipod species already mentioned and *Diaptomus shoshone* were abundant; the dominant herbivores were *D. sicilis* and the cladoceran *Holopedium gibberum*. And in the plankton a new associate appeared abundantly: a rolling, spherical colonial rotifer or "wheel animalcule" (*Conochilus*).

From here they rode a dozen kilometers (7 mi.) along the base of the Red Mountains to Heart Lake, near the foot of Mount Sheridan, where this lake received roaring streams fed by melting snow from the upper slopes. "The lake is a gem of beauty," Forbes wrote, "one of the most attractive in the Park." Similar in size to Lewis Lake, Heart was also fed by the warm springs' water of Witch Creek, but was almost twice as deep, with soundings to 45 m (148 ft.). Heart Lake had many fish—cutthroat trout, chubs, suckers, and minnows—that they collected by sieving or setting trammel nets for hours at a time. After five days of continuous work they had enough material for their first comparison with fishless lakes.

Not a single specimen of amphipod appeared in their collections from

Heart Lake, nor did any specimens of the large *D. shoshone*, both so very abundant in the two fishless lakes. What were abundant were the predaceous cladoceran *Leptodora* and the herbivore *D. sicilis* accompanying an occasional *Bosmina* and *Daphnia*.

By now Forbes was tanned, had grown a beard, and slept like a baby after days afield. He had not yet seen a bear, but there were plenty of tracks. All in all, it reminded him of his old Army days, but without the shooting. He felt refreshed. His hectic schedule at home rarely entered his mind. Rather than just the two weeks they had been in the park, it seemed like two months. His only gripe was some indigestion that he thought came from the sourdough bread; his guide said it was a common ailment on the trail.

On July 31 they moved north to the west bay of Yellowstone Lake near Upper Geyser Basin, where the lake received their usual treatment, and Linton shot fish-eating water birds for his study of trout parasites. This was the big lake, 22 × 32 km (14 × 20 mi.), with 160 km (99 mi.) of shoreline. After five days they moved to the north end of the lake at the outlet and to the Yellowstone River for another week of collecting.

At Yellowstone Lake the two amphipod species and *Diaptomus shoshone* were abundant but not to the extent observed in smaller fishless lakes. The only fish in this lake was the native cutthroat trout, *Salmo clarkii*. (Forbes called it *S. mykiss*.) In open water, whether deep or shallow, they found enormous numbers of *Conochilus*, the gelatinous colonial rotifer. The most common crustacean in the plankton by far was a new species of herbivorous cladoceran, *Daphnia pulicaria*, followed by *Diaptomus sicilis*, and *D. lintoni*.

The last lake for comparison is fishless Grebe Lake, smallest of the five lakes. Forbes visited here deep in the forest on August 22 at the headwaters of the Gibbon River, northwest of the big lake. Although turbid, Grebe showed a faunal list surprisingly similar to that of the larger, clear Shoshone: the two very abundant amphipod species along with *Diaptomus shoshone*, the herbivore *D. lintoni*, and very few *Daphnia*.[5]

To summarize, in fishless lakes (Shoshone, Lewis, and Grebe) the dominant predators on zooplankton were the amphipods and the large copepod, *Diaptomus shoshone*. The dominant crustacean herbivores here were the two smaller species of *Diaptomus* copepods, *sicilis* and *lintoni*, and the cladoceran *Holopedium gibberum*. In Yellowstone Lake native cutthroat trout was the dominant predator on zooplankton, and the invertebrate predators were exactly the same as those in fishless lakes but less abundant. Here the dominant herbivore was the cladoceran *Daphnia pulicaria*. The colonial rotifer, *Conochilus*, occurred abundantly with *Diaptomus shoshone* in two of these four lakes, but not when that predaceous copepod was superabundant.

In Heart Lake, cutthroat trout, chubs, suckers, and minnows were the predators on zooplankton; the cladoceran *Leptodora* was the dominant invertebrate
predator. The predaceous amphipods and the large copepod *D. shoshone* were
absent. Here the dominant crustacean herbivore was again *Diaptomus sicilis*.
Once again Forbes had identified the probable roles of both predation and
competition among animal populations.

In his 1893 Yellowstone Park paper Forbes mentioned that the fishless
Shoshone and Lewis Lakes had "since been stocked with trout by the U.S.
Fish Commission." If so, rainbow or brook trout, or both, were probably the
species. The effect of stocked trout on the structure of invertebrate communities of these pristine lakes does not seem to have been followed up on until
the second half of the twentieth century, when thorough studies of this kind
were made elsewhere in North America and in Europe.

After another week of collecting in the northwest quarter of the park,
Forbes's party stopped work on August 30 and went chasing butterflies and
picking flowers. In all respects it had been a successful trip. Forbes had due
praise for his resourceful guide, Elwood Hofer, and acting superintendent of
the park, Capt. F. A. Boutelle, U.S. Army, for his indomitable aid, especially in
providing Forbes with an army horse after his first mount proved not up to
standard.

This trip brought Forbes a world of good. Rejuvenated, he sailed through
the following academic year, and early in the summer of 1891 he responded
eagerly to a letter from Richard Rathbun asking if he wished to return again
for further fieldwork. The ichthyologist Barton W. Evermann would be working in the headwaters of the Missouri and Columbia Rivers. He sought out
localities appropriate for new trout hatcheries, and Forbes was welcome to do
the invertebrate work at commission expense. Forbes picked a field assistant
from his laboratory staff, H. S. Brode, who the previous summer had been a
student at Woods Hole Marine Biological Laboratory. Forbes departed on
August 10 after writing Rathbun that he had made contact and would "try to
hook up with Evermann" in the field, which he did later that month.

Forbes spent nearly 40 percent of his three and a half weeks in the field at
fish-bearing Flathead and Swan Lakes, north of Missoula, Montana, and in
various streams and rivers nearby. Flathead Lake lay at 904 m (2,965 ft.) in a
troughlike valley with the Mission Range on the east. Only about a third as
high in elevation as Yellowstone Lake, Flathead Lake at 27 × 38 km (17 × 24
mi.) was larger in area, but not as deep. Swan Lake, only 13 km (8 mi.) away to
the east, pointed like a silver finger 1.6 × 19 km (1 × 12 mi.) long between the
Mission and Kootenai Ranges, rugged peaks shining with brilliant snow-filled
ravines that long-ago summer.

On Flathead Lake, Forbes stayed at the Helena Rod and Gun Club, and he collected with the club's small steam launch. They rode to Swan Lake by saddle horse. At both lakes the amphipods *Gammarus lacustris* and *Hyalella azteca* were present. *Daphnia thorata*, another cladoceran species new to science was the most prominent zooplankton herbivore. This species along with the three from Yellowstone Park—*Daphnia pulicaria*, and the two diaptomid copepods, *lintoni* and *shoshone*—have all stood the test of time as valid new species and alone made Forbes's western work worthwhile.

A few days before leaving for Illinois in September, Forbes sat before a fire at twilight at Yellowstone Lake, where he had finished up some dredging not completed the year before. He wrote Clara that he hoped again to bring back "a fresher mind and heart" for his work at home. He would need both these and much more for what lay ahead in the coming eighteen months.[6]

That fall Forbes was asked by the Illinois Board of World's Fair Commissioners to produce a plan for an exhibit on work of both the state laboratory and the entomologists' office for the 1893 World's Columbian Exposition in Chicago. Forbes's laboratory and office staff soon doubled to fifteen, with taxidermists, an artist, a secretary, a librarian, and entomological, ichthyological, and zoological assistants—all paid by the commission. Collections made before the exposition and studies generated by the staff's preparations materially advanced the natural history survey of Illinois, equivalent, Forbes thought, "to the product of five years of our ordinary operations."

During 1892 Forbes received notice from the U.S. Fish Commission concerning possible further support for his aquatic work on Lake Michigan with the vessel *Fish Hawk*. He also was granted federal funds for work that summer on Wisconsin lakes. He learned more too during a July trip to Washington about the National Aquarial Exhibit for the Columbian Exposition and a critical need for new leadership for this stalled project that Fish Commissioner McDonald spoke for most forcefully. Yet Forbes wanted counsel from colleagues regarding reports of internal dissension, not only about the project, but about McDonald himself. Forbes knew Colonel McDonald was not scientifically trained but found him open to putting the fisheries on a scientific foundation. Some, like A. E. Verrill of Yale, now personally disliked McDonald and told Forbes so, despite Verrill having earlier supported McDonald for the commissioner's position. In a letter to Forbes in late October, McDonald reported insubordination by the current director of the project, a Mr. Collins, and personal attacks on himself by Verrill. After emphasizing the importance of the project, he then asked, "is it not possible that you could . . . take the position of Director of the aquarial exhibit?" and asked Forbes to meet with him in Chicago. He would perhaps be speaking with President Cleveland about the subject and said that Forbes could aid him greatly in getting this matter clarified. After their meeting in Chicago, Forbes agreed to become director as well as assistant to the Board of Management of the U.S. Government Exhibits. His salary would be $2,400 dating from January 1, 1893, increasing to $3,000 by May 1, when the exposition opened, and continuing through October 31.

Forbes's acceptance of this post came on the heels of his election as dean of the new College of Science. His stature now was such that forbearance was required by officials at the university for his many duties. The aquarium project, he said, would absorb much of his time and energy for ten months, carry him away entirely from the university except for an occasional day, and find him periodically out of state. As if that were not enough, Forbes had also been delegated chairman for the International Congress of Zoologists to be held at the exposition for several days that summer. General and sectional sessions would cover topics on animal morphology, physiology, and psychology, on zoogeography, philosophical zoology, and the state and progress of zoology in the world. Clearly Forbes's abilities and scientific reputation were now widely recognized and rewarded. He reported as much to Clara after a pleasant visit with his colleague Dr. Charles Whitman, head of zoology at the new University of Chicago and director of the Marine Biological Laboratory at Woods Hole. In terms of ecological sophistication, Forbes in the late 1800s was among the top half-dozen biologists in the United States and was well-known among similarly oriented biological scientists in the world.[7]

When Forbes took charge of the aquarium annex of the new Exposition Fisheries Building on January 1, 1893, it was not a pretty sight: unfinished leaky tanks, no heat, burst water pipes—"a discouraging wreck," Forbes said. They had miles to go and in only four months. But he and his work force of 15–20 men did it: repairing, finishing, furnishing, and stocking an aquarium complex of 140,000-gallon capacity, despite a winter and spring with incessant storms.

The circular aquarium annex measured 40 m (131 ft.) in internal diameter, and easily held fifty tanks placed in two concentric circles. Within the inner circle lay a central pool with a white sand bottom, and 6,000 additional gallons of fresh water, 0.8 m (2.6 ft.) deep and 8 m (26 ft.) in diameter that Forbes used for receiving freshwater fish and as a home for larger species. Tanks and pool were cleverly lighted through side and roof windows, with accessory electric light supply. Freshwater was cycled from Lake Michigan, filtered, and aerated in tanks, and returned to the lake. Saltwater was obtained from the sea off Beaufort, North Carolina, and sent by railroad tank cars to Chicago, where it was held and cycled to and from fifteen tanks of 40,000-gallon capacity via an underground cistern and a pressurized tank on the roof of the Fisheries Building. Fish were fed minnows, minced liver and beef, and chopped clams.

Collections of fish and invertebrates were made by sending men and special railroad tank cars to locations on the Illinois and Mississippi Rivers, to U.S. Fish Commission stations in Colorado, Missouri, Maine, Michigan, and Washington; to Lake Erie, North Carolina, Woods Hole, Florida, and California; and to state fish commissions in Iowa and New York. Forbes chose freshwater fish species to illustrate those of lakes, rivers, and streams of the interior United States, and marine fish and invertebrates for popular interest. Displays of fish were grouped by families or by species that lived together naturally. Total species represented during the exposition were 115 freshwater and 93 marine, 39 of the latter being invertebrates.

The Fisheries Building was designed in Spanish Romanesque style by Henry Ives Cobb, master planner for the University of Chicago. In dull brown stone, and with an iron roof colored to represent tile, the building was more than two and a half football fields long, covering two acres. The central section held general fisheries exhibits, the north annex an angling exhibit, and the south annex the aquarium. The building stood at the northeast corner of the lagoon in Jackson Park at the junction of the lagoon and north inlet from Lake Michigan, directly across the inlet from the U.S. Government Building to the south, where Forbes had his office. These buildings and 218 others made up what some called the erection of a "White City from a swamp."

By mid-February 1893 a busy but lonely Forbes wrote his wife to "please come to Chicago." Clara came and brought with her Richard, age 6, arriving in Chicago on March 7. Four days later Richard fell ill with a sore throat and very difficult breathing; a yellowish membrane appeared on his pharynx indicating probable diphtheria. Dr. Bayard Holmes performed an intubation, inserting a tube in the larynx to assist breathing. The boy was then fairly comfortable and for 4–5 days rested and seemed to be recovering. But on the sixteenth he worsened in great pain. Dr. Holmes gave morphine, and Forbes, who was not in Chicago, was called at once, arriving early on the eighteenth. By now Richard was exhausted and wanting to go home. Sedated, he slept off and on holding his parents' hands, and Clara said, "fought a brave fight." Calling for water and slipping away the Forbes's youngest child died that Saturday evening. Forbes never again spoke in public of his son.

In despair they were told that Richard had to be buried in Chicago to prevent spread of the disease. He could not be brought home until fall. His grave lay in Hyde Park just north of the exposition grounds. Meanwhile the Forbes burned their excess clothes, disinfected their needed clothes, themselves, and their belongings, and fumigated their hotel room. Clara left for home soon after.

During April as aquarium preparations heightened, Forbes allowed himself to grieve and took solace from a Lieutenant Hoppin, U.S. Army, "whose words of manly sympathy were like a new dressing to an old wound." His aquarium work was unending: overseeing the water supplies, electric power, air compressor, fish collections, contractors accounts, foreign fish commissioners, visiting bureaucrats, and myriad other details, but they seemed to take his mind off Richard. Not working brought back the pain. But then he reflected that now at least he had a home, wife, and family to go back to, unlike the "homeless wandering of 25 years ago, where my future was all a cloudy blank." To keep Clara's image before him, one day he took a long walk to the southern end of the exposition grounds. There he visited the site of the old hotel at lakeside where he stayed twenty years ago after visiting Clara in La Porte, Indiana, before they were married. It brought back that "day in heaven" he felt after seeing her again.

"You will come up here soon for little while," he now wrote Clara in Urbana. "My first grief has taken as complete possession of me as did my first love—but I shall rally soon." And a few days later: "I feel that the Aquarium is sure to be one of the great attractions of the Exposition."

May 1, 1893, dawned cold and misty with a threat of rain at Jackson Park, but the sun broke out just before noon. President Cleveland opened the exposition just after noon by pressing a special gilded telegraph key that completed the circuit and set the fair in motion. Earlier that morning a parade made its way down Michigan Avenue led by mounted Chicago hussars, park police, Illinois national guardsmen, and the U.S. Cavalry in dress uniform, followed by VIP's in twenty-two carriages pulled by high-stepping horses.

That afternoon the aquarium staff was ready with eleven carloads of freshwater and marine fish in their aquaria in preparation for visitors who paid 50¢ for admission to a fair with some 65,000 exhibits. As Forbes predicted, the aquarium "was regarded as one of the most attractive features" of the fair, "a source of wonder and delight for months to several million people." Overall almost 21.5 million people passed through the fair turnstiles. Considering repeat visitors, estimates were made that 5–10 percent of the U.S. population saw the fair.

Meanwhile, Forbes found all of his nonaquarium duties in dire need of attention. With the aquarium up and running he requested a new administrative arrangement from Commissioner McDonald. Not wanting to lose Forbes as director, but understanding his situation, McDonald suggested the appointment of an executive officer to handle day-to-day details of administration, accounts, and maintenance, allowing Forbes to concentrate only on the overall plan and performance of the exhibit.

Both men had hoped that the aquarium might be made permanent in Chicago after the fair, that with the addition of a biological laboratory and a federal fish cultural station it might even, McDonald wrote, "take rank with the aquaria at Brighton, Berlin, and Naples." Offers were made of the aquarium contents to the trustees of a proposed Columbian Museum, and later to the South Park Commissioners of Chicago. The federal supervisor of the aquarium, his assistants, and even Forbes were all willing to stay temporarily until transfer of contents and control were made for a new public aquarium. But all to no avail, and in early November 1893 Forbes directed the specimens to be sacrificed, preserved, and distributed for use of the state laboratory, and the colleges and high schools of Chicago.

The past dozen years had been productive and stimulating for Forbes's investigations into aquatic systems from both a scientific and an applied point of view. His state laboratory and its work had benefited from wide public exposure and by Forbes and his staff's participation in several research projects involving public waters that received both state and federal support. Forbes was now poised for a new and expanded initiative in aquatic biology that would engage and challenge him for the rest of his life.[8]

Forbes farm, 140 acres, 1844–61. Ridott, Illinois. Courtesy Forbes family.

Private Stephen Forbes, age 17, early 1862. Courtesy Richard Forbes.

Major Henry Forbes, age 31, 1864. Courtesy Richard Forbes.

Major General Edward Hatch, 1867, former commander Fifth Division, Wilson's Cavalry Corps, Military Division of the Mississippi, 1864–65. Courtesy National Archives.

Lieutenant General Nathan Bedford Forrest, CSA, 1865. Courtesy National Archives.

Nettie Forbes, age 20, 1865. Courtesy Forbes family.

Rush Medical College, 1855–67, Chicago. Courtesy Chicago Historical Society.

Illinois Normal University, 1880. Courtesy Illinois Natural History Survey.

Agnes Van Hoesen Forbes, age 61, 1861. Courtesy Forbes family.

Stephen Forbes, age 36, 1880. Courtesy Richard Forbes.

Clara Gaston Forbes, age 32, 1880. Courtesy Mary Scott Cox.

Fisheries Building, World's Columbian Exposition, 1893, Aquarium Annex at left. Courtesy Chicago Historical Society.

Quiver Lake, Illinois River from the east bluff near the Illinois Biological Station headquarters, 1896. Courtesy Illinois Natural History Survey.

Illinois Biological Station, floating laboratory with steamer Illini *alongside, 1896. Courtesy Illinois Natural History Survey.*

Interior of floating laboratory, 1896. Courtesy Illinois Natural History Survey.

Orchard spraying operation with whale oil soap. Notice poles with spray nozzles attached by hoses to spray tank, 1898. Courtesy Illinois Natural History Survey.

Unprotected cornfield (left) and protected cornfield (right). Note dark line of creosote on the brow of the ridge made by a ploughed furrow between the fields. Chinch bugs are turned away as they travel on foot up the furrow. Courtesy Illinois Natural History Survey.

Forbes entomological laboratory, University of Illinois, 1900. Courtesy Illinois Natural History Survey.

Robert Richardson (left) and Henry Allen, Illinois River, 1910. Courtesy Illinois Natural History Survey.

Stephen Forbes, age 71, 1915. Courtesy Richard Forbes.

Stephen, age 77, and Clara Forbes, age 73, at Dunewood, Michigan, 1921. Courtesy Mary Scott Cox.

Stephen Forbes, age 80, 1924. Courtesy Richard Forbes.

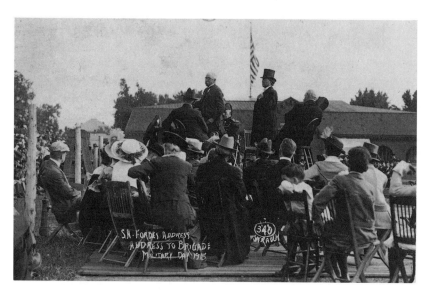

Stephen Forbes's (hatless, center) *address to Army ROTC Brigade, Military Day, University of Illinois, 1915. General Leonard Wood, U.S. Army, behind Forbes; President Edmund James with high hat. Courtesy Illinois Natural History Survey.*

CHAPTER 8

The Illinois Biological Station, 1894–1915

I
n 1894 Forbes and his colleagues had lighter than usual teaching duties because of curricula changes in the zoology department and in the new College of Science. This left more time for research ardently encouraged by Acting Regent Thomas Burrill. Wanting to infuse new life into his research program, in early March Forbes requested $2,000 from the board of trustees to establish an aquatic biological station on the Illinois River "for continuous scientific investigation." The station would address the biology and ecology of a great river system "in interior North America"—an untouched subject in the United States during a time when marine stations were more common both here and abroad. And as he would say after his studies were under way, he found rivers more interesting than lakes: "a lake has an aspect, a constitution; but a river has a character, a behavior."

Like Burrill, Forbes argued for the university to produce more knowledge as well as passing it on—pure science, yes, but also attention to potential application of findings to economic problems and the public welfare. For example, Forbes mentioned "the improvement of fish culture, the prevention of a progressive pollution of our streams and lakes," and better awareness of Illinois's natural resources. If this station was well-supported, he wrote, the University of Illinois would be better equipped for freshwater research and study in ecology "than any other institution in America." In fact, Forbes reported to the trustees, it was his intent to eventually extend the work to the "aquatic life in the Mississippi Valley in all of its relations, scientific and economic." For now, though, studies would focus on the effect of the "periodical overflow and gradual recession" of water in the Illinois River flood plain on aquatic

143

plants and animals. Coincidentally he would compare the biological and chemical features of the river before and after the opening of the Chicago Drainage Canal, currently under construction for carrying increased sewage and other wastes southward into the Illinois. Forbes had here a clear blueprint for pioneer American studies in environmental science.

The station would also serve as an important adjunct for students of botany and zoology. For years Forbes had deplored both the bookworm and the "laboratory grub." "Both have suffered, and almost equally, from a lack of opportunity to study nature alive." A limited number of advanced students and independent investigators would therefore be welcome to pursue their studies at the station through the summer school begun by the university in 1894. In this fashion, too, the main focus of the station would be investigation.

This was an opportune time for Forbes to propose a new scientific initiative given the past three years of reform, strengthening, and general transformation of the university under Burrill. Forbes received most of the money he requested, but funds would not be available until July 1, so Forbes opened the station April 1 with state laboratory money and temporary leased quarters in Havana, south of Peoria, on the east bank of the Illinois River. Havana, an old riverboat and fishing town, was headquarters of the Illinois State Fish Commission and had good rail and road connections. Forbes also rented a cabin boat sleeping four and used it as field headquarters at Quiver Lake, just north of town. A pollution-free springwater supply, fine campground, and excellent river and bottomland habitats made the location perfect for field studies on diverse populations of aquatic plants and animals. Frank Smith, instructor of zoology, would serve as station superintendent during 1894–95, after which Forbes received appropriations of $2,500 for equipment and $6,000 for operating expenses during 1895–97; similar funding continued through the following decade. During these years Forbes's biennial state appropriations for all his investigative work including the station at Havana ranged from $25,300 to more than $36,000, testimony to the influence and respect Forbes commanded with the university faculty and administration as well as the state legislature in his efforts to develop an institutional basis for ecological research in Illinois.

Yet Forbes took nothing for granted. During 1894–96, when vital support was needed, he cultivated visits and correspondence with colleagues and relevant journals, from which he gathered their comments regarding the potential value of his fledgling biological station. He received input from Charles Whitman, David Starr Jordan, Edward A. Birge, University of Michigan zoologist Jacob E. Reighard, Brown University zoologist Alpheus S. Packard, ichthyologist George B. Goode (director of the U.S. National Museum), the journal

Natural Science, U.S. Fish Commissioner Marshall McDonald, and others from abroad. Statements were sent to new University of Illinois president Andrew S. Draper and the board of trustees. Colonel McDonald wrote Forbes, "The investigation you contemplate is absolutely necessary as a foundation upon which to build an intelligent, rational administration of our fishery interests. A knowledge of life in relation to environment is an important subject which biological investigators have not heretofore sufficiently dealt with. . . . You may count on the Commission for any cooperation and aid that we may be able to give you."[1]

"Illinois is essentially a river state," Forbes often said. The state featured 6,400 km (3,968 mi.) of navigable rivers, including the Mississippi, Ohio, and Wabash surrounding it on three-quarters of its borders. The Illinois River, an interior sluggish lakelike stream, flows 432 km (268 mi.) from northeast to southwest within a basin of 75,110 sq. km (29,000 sq. mi.) or 43 percent the area of the state (fig. 5). At Havana a century ago the river expanded 6.4 to 8 km (4–5 mi.) wide at flood, and shrunk to less than 155 m (508 ft.) wide at low stage, leaving as it fell, lakes and backwater sloughs occasionally filling up with choking aquatic vegetation. Forbes described the scene there in 1896:

> The forest itself, beginning at water's edge with a billowy belt of pale green willows, is an untamed tract of primitive wilderness. Elms and pecans and sycamores tower overhead. The shallow lakes and swamps are glorious in their season with the American lotus and the white water lily. Waterfowl abound, and fish lie in the shallows, basking in the summer sun. The microscopic life in a cubic meter of water is at certain times far in excess of the amount recorded for any other situation in the world.

Numerous research topics occurred to him: systematics, life histories (reproduction, growth, and survival), physiology, and behavior, but the overarching tie was the newly marked ecological view that Forbes had followed instinctively for twenty years and now eagerly embraced in a thoroughly conscious way. It would lead their efforts. He had no illusions that this would be easy. Surely the entire river, or even a large part of it, could not be studied all at once. Specific findings along the way, however, would be used practically and generalizations made and modified as their knowledge grew. Continuous observations, erection of hypotheses, and experimental procedures would necessarily be employed. Both field and laboratory work were mandatory. "It is, in fact," Forbes wrote, "the biological station, wisely and liberally managed, which is to restore to us what was best in the naturalist of the old school, united to what is best in the laboratory student of the new." Optimistically, Forbes hoped the station would do for aquatic biology and fish culture what

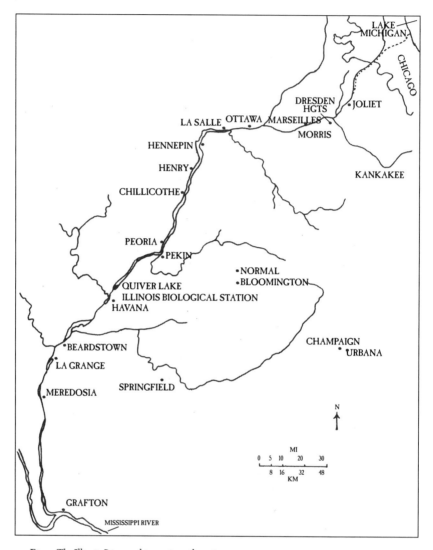

Fig. 5. The Illinois River and its major tributaries.

the agricultural experiment station was doing for agriculture and economic entomology.

His initial staff in 1894 was small but experienced and effective: Frank Smith, plankton and freshwater worms; Charles Hart, insects and mollusks; Adolph Hempel, protozoa and rotifers; Thomas Burrill, aquatic plant consultant; and Stephen's son Ernest (B.S. in zoology, Illinois, 1897), general collector during college vacations. Miles Newberry signed on as boatman and me-

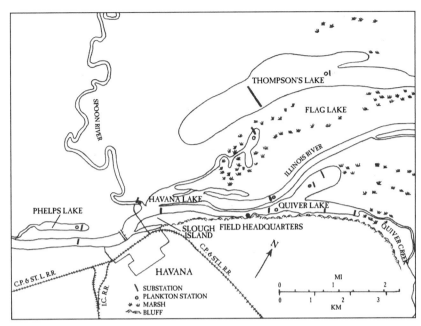

Fig. 6. Field of operations of the Illinois Biological Station near Havana on the Illinois River at low water, 1898. (Adapted from pl. 1 in S.A. Forbes [1898], Biennial report of the director for 1897–98, Illinois State Laboratory of Natural History, Urbana.)

chanic, and Professor Arthur Palmer, of the university's Department of Chemistry, began water analyses that May.

Their substations (collection points) generally numbered seven at first and later doubled, covering locations in the main river, sloughs and lakes, and some creeks, all within a radius of five kilometers from the cabin boat at Quiver Lake (fig. 6). They collected at 1–3 week intervals during April—October, and three-week intervals from November to March.

At midriver and midlake substations, surface and bottom collections were made with tow net and dredge. Plankton was collected quantitatively by hauling a #20 plankton net, attached to a small wheeled carriage, along a tightly stretched line running obliquely from bottom to water surface over a known distance and rate of speed. This was done because most substations were too shallow for satisfactory vertical hauls. They also made fish seine, hand-net, small dredge, and hand collections along shorelines, and they took water temperature measurements daily. With some modifications for plankton to be described, this was their general collection scheme followed in the Havana area from 1894 to mid-1896.

During the first summer of operations, the leased cabin boat proved too small and uncomfortably hot for practical use. They needed a special house-boat built for their work, one made usable by being movable. The solution was a 6 × 18 meter (20 × 59 ft.) houseboat for staff and fifteen other researchers towed by a 2 × 8 meter (7 × 26 ft.) steam launch seating seventeen. Built with Washington fir and Georgia pine, the houseboat was cleverly ventilated by Forbes's design and remained very comfortable. It featured four rooms: a large laboratory, an office-library, a kitchen, and a storage room. Launched in early 1896, the houseboat remained in use until replaced by another in 1935 and then by the first permanent land-based laboratory in 1940. As Forbes put it, during his watch "the Illinois Biological Station . . . was never really stationary, being actually portable and afloat, its laboratory and store-houses being boats in-stead of buildings." Although unique for the United States, Forbes's biological station was predated by two European freshwater stations: one in Bohemia, where Dr. A. Frisch also had a portable zoological station, and another at Plön, Holstein (Prussia), directed by Dr. O. Zacharias.[2]

No sooner had the biological station started up than Forbes again had his feet put to the fire. His professed interest in U.S. Fish Commission work as well as water quality and pollution brought a request from Rathbun and Mc-Donald to look into damage to food fish on the Wabash River from oil pollu-tion. Congress had asked for an investigation. Forbes spent several days on the river near the source of pollution at Terre Haute, Indiana, and downstream on the Illinois side talking with fishermen, merchants, and town officials. Pollu-tion by oil from 2–3-year-old wells from an illuminating gas works, but pri-marily from a fuel gas works (several barrels or more per day), had injured fish up to 64 km (40 mi.) downstream, crippling the fishery and the sale of fish. Fortunately, the fuel gas executive showed Forbes his works and pledged coopera-tion in solving the problem. Professor A. N. Talbot, sanitary engineer from the University of Illinois, recommended that oil and tarry wastes were easily sepa-rable from wastewater by subsidence in a large tank at relatively little expense.

Forbes had been impressed with the volume and diversity of plankton or-ganisms collected in the station's first year of operation. Consequently, he was delighted to learn that a new Ph.D. from Harvard, Charles Kofoid, was filling in for Professor Jacob Reighard and teaching at the University of Michigan, and was also working on quantitative plankton estimates at Lake St. Clair, Michigan. Kofoid had recently contacted Forbes about a possible position with him at the recommendation of Birge at Wisconsin, who could only offer Kofoid a one-year post. Forbes soon offered Kofoid the positions of planktonologist and superintendent of the Illinois Biological Station begin-ning July 1, 1895, which he accepted.

In addition to his classic morphological work (on mollusks) for the doctorate, Kofoid had an instinct for the ecological point of view and was interested in fish and their parasites, besides plankton. Most important, he brought to Illinois his scientific acumen, his thorough well-planned sampling, and quantitative estimates for sampling error. These skills were decisive for his overall objectives planned with Forbes: species composition and abundance of river plankton over time; and correlations of these with water temperature, chemistry, and biological productivity of the river.[3]

Because of clogging as well as large losses of small phytoplankton cells going through typical plankton nets (50 μm mesh) even in clear water, Kofoid came up with a hand-pump for sampling. This took water from a specified depth, then through a fine net and filter on board a boat, and finally into a centrifuge to capture minute species. Kofoid used this method for three years on a weekly or fortnightly basis at six substations in the Illinois River. The tiny phytoplankton he collected (<20 μm) forged a "fundamental link," Kofoid said, in the aquatic environment. Soon the German zoologist Hans Lohmann, who had seen these "nannoplankton" in gut contents of ocean plankton animals, set out to assess them using Kofoid's method and described their importance in the Baltic Sea and elsewhere.

Together Forbes and Kofoid monitored Illinois River water chemistry, with samples analyzed in Arthur Palmer's chemistry laboratory on campus. The number of samples they generated threatened to overwhelm his facilities, but the state legislature came to the rescue with a total of $8,000 during 1895–96 and a directive for the university trustees to establish a State Water Survey through Palmer's laboratory. Plankton and other operations at the Havana station yielded water samples analyzed for dissolved oxygen, carbon dioxide, free and bound ammonia, nitrites, nitrates, and chlorine.

In late 1896 Forbes reported good progress with the station's work but had concerns about cost overruns for equipment and boat machinery. Bills were paid, but Forbes scaled back all operations except Kofoid's. By 1896 station work had accumulated more than six thousand collections of various kinds, resulted in a dozen papers published or in press, and supported the first Summer School of Biology at Havana. A class of seventeen instructors of biology from high schools and colleges included a 22-year-old Charles C. Adams, an early animal ecologist who published the first *Guide to the Study of Animal Ecology* (1913), and 27-year-old James G. Needham, who studied aquatic insects and general limnology and later became an eminent specialist on systematics and ecology of dragonflies. Forbes's daughter Bertha, now a high school biology teacher, and four other students were also residents of an informal summer program at the station during the summer of 1897.[4]

Meanwhile, the mid-to-late 1890s were not kind to U.S. Commissioners of Fish and Fisheries, nor to nominees for the position. Colonel McDonald died in 1895, and although his associate, Richard Rathbun, received strong support from Forbes, Jordan, and others, George Brice, a former navy captain was appointed commissioner in 1896 by President Cleveland. Then in spring 1897 a groundswell of support for Forbes's nomination as commissioner came from scientific colleagues and the first district congressman from Illinois to President McKinley. But in 1898 the president appointed George M. Bowers of West Virginia in return for his political support. Forbes would probably not have accepted the position in any event. Nevertheless, scientists were placated by the presence of Dr. Hugh Smith, now director of the commission's scientific work. That July, Forbes celebrated his own independence while he searched in northern Illinois for the newly arrived San José scale and took a bicycle trip in his "native land" between Rockford and Freeport. Under a glorious summer sky he drank in the prairie and wrote Clara that he looked forward to seeing "the old place and stone schoolhouse once more."[5]

Although Forbes had worked on fishes for almost twenty-five years, his systematic and ecological studies with them had been intermittent and more recently subordinate to his entomological duties. This matter was redressed beginning in 1898 through a collecting effort with a view toward an elaborate report—eventually a book—on the fishes of Illinois. He was assisted by Wallace Craig (on the Illinois and Spoon Rivers with connected creeks), Thomas Large and others (on eastern, southern, and southeastern Illinois rivers and streams), and two dozen principals, teachers, superintendents, professors, and citizens throughout the state, whose many wagon, boat, and foot trips yielded specimens from several hundred Illinois localities. Forbes then augmented these specimens by towing his station houseboat 70 km (43 mi.) south to Meredosia (see fig. 5) for fish collections during 1899–1900.

Always on the lookout for better and more efficient use of resources, Forbes now came up with a win-win recommendation approved by the president and trustees. State laboratory staff would be assigned by him when needed—at regular pay—for service at the biological station or entomologist's office. The same people would be available to university science departments for teaching "with approval of all concerned," provided there was a mutual advantage to both the state laboratory and university. The spirit of this recommendation had come from his experience with his manifold operations, including the Columbian Exposition. His successes had reflected equally well on the university, the state-at-large, and on his scientific reputation.[6]

As the nineteenth century wound down, Kofoid finished his five-year study of the Illinois River plankton. Forbes reported that no other study at the station to that time had attracted "more general attention among scientific men or is likely to lead to more interesting and important results." Over an eleven-year period from 1897 to 1908, Kofoid would report his findings in almost 1,000 pages of elegant scientific sleuthing. It was the largest and first comprehensive study of river plankton ever made and set the tone of station operations for years.

The new plankton method generally yielded plankton volumes three times those captured with the usual silk net alone. Hence a yearly average plankton sample Kofoid took in the main channel of the Illinois River at Havana in 1898 caught eighteen phytoplankton cells for each zooplankton individual. Zooplankton—mostly protozoans, rotifers, and entomostracans—made up less than 6 percent of all plankton organisms, whereas more than one hundred species of Protozoa made up 98 percent of the zooplankton. Kofoid also made preliminary mean annual production estimates of plankton and calcu-lated that each year the aggregate weight of plankton passing downstream reached fifteen times the weight of fish taken from the Illinois River.

Quiet lakes (lakelike expansions connected to the river) and backwaters of the Illinois generally displayed even more abundant plankton than the flowing river water; outflow of these lakes and backwaters therefore enriched the main river. Lakes with thick vegetation, however, had the least amount of plank-ton—partly due no doubt to obstructed light.

Kofoid readily characterized a winter plankton assemblage with a small number of species peculiar to it, along with some perennial types. The sum-mer plankton displayed a much larger number of species common to it along with perennial ones. Plankton of the Illinois River included 528 common and recurring species and varieties.

Kofoid left Illinois and went on to a distinguished career as a professor in the zoology department of the University of California at Berkeley (1901–36), and as a research scientist and assistant director at the Scripps Institution of Oceanography (1910–23). His experiences and work with Forbes on the Illi-nois River plankton materially influenced his research program over the rest of his life: marine plankton, the morphology and taxonomy of the Protozoa, as well as marine zoology, parasitology, and tropical medicine (amoebiasis). He was elected to the National Academy of Sciences in 1922.[7]

During that decade straddling the turn of the century, when Forbes entered his sixties, his personal work in aquatic biology was minimal—he had no pub-lished papers on the subject from 1895 through 1904. He was, however, fully occupied with his deanship, San José scale, and corn insects. He also directed Frank Smith that fish collections be continued, with the station houseboat moved 265 km (164 mi.) north to Ottawa for work during 1901–2, and then 50 km (31 mi.) south to Henry (for this and Ottowa, see fig. 5) in 1903. For several years Forbes's valuable station artist, Lydia Hart, had sketched exquisite colored drawings of living fish kept in the houseboat aquaria. Forbes's intention was to include colored plates of each Illinois fish species in the comprehensive treatise on fishes being prepared. These drawings were continually checked and corrected for true color, form, and attitude.

By now Forbes was a bit heavier and more gray, but still vigorous—he rode a bicycle longer than most of his contemporaries and sometimes even took a brisk evening jog. He thoroughly enjoyed the theater, wrote poetry, and was, said his son Ernest, "in all ways a man of refined sensibility." Disciplined yet "full of sentiment," wrote his daughter Ethel, "he was fearful of seeming sentimental." Usually reserved, he still retained from boyhood a measure of shyness, and to strangers, sometimes appeared brusque with his drooping moustache and prominent eyebrows. Having made his distinctive mark, he was now more mellow and patient and purposely helpful to younger colleagues. Still a stickler for clear and careful writing, he held a high standard for his staff and was not easily satisfied. His reports, letters, and scientific papers generated glowing comments from colleagues and correspondents for their comprehensiveness, excellence, utility, and beauty. Few people could be in Forbes's presence for more than five minutes, his friends thought, without sensing his keen inclination for vigorous scholarly work.

Typically, Forbes was lively and often restless. Ernest wrote that "his course through the Natural History Building could be traced by the slamming of doors behind him. The attitude of command . . . was habitual, and he expected action on the part of his subordinates." Yet, among those colleagues, friends, and family of whom he disapproved, he was fair and temperate in his treatment and expression of them. Ernest wrote that he was "too wise to hate anyone," but "on occasion he could demonstrate a power of description equivalent to vivisection." Not fond of contentious people or situations, yet if necessary, his command of the spoken and written word was frank and just for the situation at hand, making him someone not to trifle with.

By the 1890s Forbes had joined the University Club, Natural Science Club, and Philosophical Club, and had founded the Golf Club. Clara and he had also joined the Theatrical Club. Moreover, she had helped found the Tuesday Tea Club that later became the University Women's Club. Forbes enjoyed kindred philosophically spirited men with whom he met in the Theory Club, as Charles Kofoid recalled:

> He was the leading spirit in the Theory Club—a group of six, consisting of Professors Kinley (later President), Palmer (Chemistry), Townsend (Mathematics), Daniels (Philosophy) and myself, the youngest of the group. We met regu-

larly in the home of some member, and digested and discussed in a heavily nicotinized atmosphere such mighty treatises as Karl Pearson's "Grammar of Science" and Ward's "Naturalism and Agnosticism." As a dialectician, Forbes was keen in analysis, alike in attack on and defense of the position of the author under discussion and of his critics among our members. We came to look upon him much as an arbiter. He was at his best when he was in the midst of the clarification of some knotty problem in the philosophical relations of science, when with uplifted cigar in hand he would discourse wittily, fluently, and pointedly, till his cigar went cold.

The Forbes family still lived in their 1209 W. Springfield Avenue home, just north of the Interurban Railroad line running through what is today the engineering campus, and across from old Illinois Field. Thoroughly rural in the early days, their neighborhoods' gardens, arbors, small orchards, and some pasture lingered into the new century even as the university began buying up land near them for expansion beginning in 1905. In the late 1890s Bertha was teaching school in Chicago, and the three younger children, Ethel, Winifred (later a professional musician), and Ernest, lived at home and joined their parents as members of the University Choral Society. Soon after 1905 all the children had left home, as Stephen and Clara settled into late adulthood and joined friends for summer vacations in West Virginia.[8]

By the fall of 1903, after almost a decade of biological station work, Forbes suspended field operations in order to prepare and publish a mountain of backlogged reports and papers on aquatic biology. He lacked adequate funds to do both. One exception was occasional plankton collections to 1909 made sequentially over the course of the Illinois, Mississippi, and Ohio Rivers, totaling more than 1,600 km (992 mi.) by steamers of the Illinois Fish Commission. Full-scale station operations would not resume until mid-1909.

The work of the biological station to 1903 naturally revolved around the required initial studies of animals and plants from the midsection of the Illinois River centered at Havana, with supplementary (primarily fish) studies further north and south. Forbes's topics were comprehensive: systematics and ecology; life histories; behavior of fishes; plankton; and water levels, temperature, and chemistry. They were investigated with reference to the periodic flooding and recession of water in the Illinois River floodplain—a previously unexamined subject. More than a dozen scientific papers had been published on the plankton, aquatic insects, worms, mollusks, leeches, ostracods, and fishes. Forbes said that their work was attracting much attention overseas, and a leading American zoologist wrote him that "the Illinois River promised to become soon the best known—because the best studied—of any river in the world."

At the turn of the century the University of Illinois served as the major entity for scientific research in the state, and Forbes's aquatic biological station on the Illinois River was now an active integral unit of the university. Costs of water analyses for the Illinois River were now assumed by the Chicago Sanitary District and the Drainage Commission. They were under increasing pressure to control pollution downstream, in the face of a greatly increased flow of diverted Lake Michigan water sending municipal, stockyard, and industrial wastes from a growing Chicago into the Illinois River. All of Forbes's work to date became even more important in providing a picture of biological and chemical conditions of the river before this pivotal event.

Lake Michigan had been used by Chicago residents for drinking water since the 1860s. With the rapid growth of the city in the last third of the nineteenth century, increased use was made of the old Illinois-Michigan canal, and in 1871 the partially reversed flow and pumping of the Chicago River, to divert polluted water away from Lake Michigan and southward into the Illinois River. Soon this proved inadequate, and construction of a much larger drainage canal had begun. Finally, on January 1, 1900, the Chicago Sanitary and Ship (or Drainage) Canal, 56 km (35 mi.) long, opened, fully reversing the flow of the Chicago River and sending almost 3,000 cu. ft./sec. of Lake Michigan water with diluted wastes into the Des Plaines River, and then into the Illinois at Dresden Heights (see fig. 5). This increased the average flow of the Illinois at Peoria 160 km (99 mi.) downstream by 35 percent, raised the midsummer level of the river at Havana almost a meter by 1909 and quickly doubled the surface area of lakes, sloughs, and marshes in the lower Illinois bottomlands. The stage was now set for considerable biological station work when Forbes's funds became available.[9]

One other chief unfinished project, of course, had to do with fishes—work begun more than a quarter-century earlier—and Forbes now needed someone to work closely with him for several years to complete it. In early 1903 he had such a man in Robert E. Richardson, 25, who had a master's degree in zoology from Illinois and a strong interest in fishes. Described by Forbes as bright and industrious, Richardson was hired as a biological assistant in the state laboratory. Despite concern over Richardson's aloofness and his sudden absence from work that winter due to a health-related problem, Forbes assigned him to fish collections in the spring at Henry on the Illinois River north of Peoria. By July, Richardson grew ambivalent over the position. He first wrote Forbes that he was dissatisfied because of his relatively low pay in a "nameless" position with little chance for recognition, and wished to stop work September 1. But then he recanted and agreed to work with his "best endeavors" for the laboratory after Forbes scolded him for "unsteadiness of purpose."

Forbes soon resolved this matter when Richardson was appointed nominal superintendent of the slumbering station, making him key man on fishes, then later on the Illinois River benthos, and eventually lauding him as one of the most productive research colleagues Forbes ever had over the next twenty-seven years.

Both Forbes and Richardson were well aware that many fish species showed different relative abundances in rivers, streams, lakes, and ponds. Even more fundamentally, fish species varied in the frequency with which they occurred with other fish species in a particular habitat or region. Yet a rigorous study of fish associations had not been made. Forbes's objective was to measure quantitatively the presence and strength of an associative bond between particular fish species and between the species and their general habitat features, for example, stream, river, lake, pond, or marsh, water movement, and bottom type. The correlation and variation of these two measures of association deserved attention as well.

For a start, Forbes and Richardson examined 13 species of darters (Etheostominae) from 1,072 collections taken throughout Illinois. The number of collections for each darter species ranged from 16 to 236, and averaged 82. Forbes then employed a coefficient of association: $\frac{ad}{bc}$, where a equals the total number of collections, b the number of collections containing one of the species to be compared, c the number of collections containing the second species, and d the number of collections containing both species. A coefficient of 1 indicated no associative relationship, that is, the probable and actual joint occurrences were equal. A coefficient of <1 suggested strongly different ecological or behavioral characteristics. Coefficients of 2 to 4 or higher would indicate fish being found together twice to four times or more frequently than chance would indicate.

Six darter species were more strongly associated with each other than with all other darter species, their coefficients averaging 2.48, whereas the weakest bond surfaced within a group of four "less typical" darter species averaging 1.36. Three of these latter species, plus three additional more weakly associated species, also displayed the most diffuse or feeble ecological preferences. On the other hand, the first-mentioned six more strongly associated species displayed high ecological uniformity, preferring smaller rivers and creeks with swift water and rocky or sandy bottoms—habitat characteristics of "typical darters." Hence similar ecological characteristics in this subfamily were confirmed, and they at least partially explained the strength of the associative bond among these species. Forbes believed the method showed promise, and he planned further refined use of it. His use of a statistical method for measuring association between species, and between species and their respective environments, or a similarity index, was not explored by ecologists until the 1920s.

In 1906, in a lecture at the University of Chicago, Forbes reviewed the ecology of freshwater fishes. He remarked about the sunfishes (*Lepomis*) that, using his coefficient of association, he found closely related species avoiding each other and associating with fish species far-removed taxonomically—the reverse of the situation with the typical darter association. He described here what would three decades later be suspected as Gause's principle or competitive exclusion. Forbes wrote, "I have supposed this to be an expression of the disadvantages of close competition between closely similar species, and of the advantage, consequently, of such differentiation in habit . . . [and] in ecological preference as would carry these natural competitors into non-competing ecological groups."

Finally after almost thirty years of painstaking, dedicated work, Forbes's magnum opus with Richardson, *The Fishes of Illinois*, was published in 1908. The work was well received and considered by many at the time as the best regional ichthyological treatise ever published in North America. It was not brought up-to-date for the state until 1979. The second edition appeared in 1920. Richardson's role was decisive for the completion and success of this book, not only for the fieldwork but for the organization and critical study of the collections and, as Forbes wrote, "for the preparation or revision of nearly all the technical descriptions" of the species.

Specimens were caught with sieves; trammel, set, and pound nets; or picked from fish markets or commercial fishery landings. This wide coverage revealed the salient position of the Illinois River basin: 85 percent of the 150 fish species in the state were represented there. A similar percentage of Illinois species had chief geographic affiliations with the fish faunas of the Mississippi, Missouri, and Ohio valleys, and with the Great Lakes basin. The volume also includes general ecological notes on species, a section on the topography and hydrography of the state, an appended series of detailed maps of Illinois showing the distribution of 98 important species, and a short description of the fisheries of Illinois. Three dozen species were marketable, a dozen of which were of fine food quality. These were now the halcyon days for the Illinois River.[10]

By 1908–10 three million people lived in the Illinois River watershed whose wastes were diluted but 96 percent untreated. At the same time fish yields from the lower river (south of Hennepin) had almost doubled since 1900, making up 10 percent of the U.S. fish catch, and assisted by a near doubling of fishermen. Since plankton volumes per cubic meter at Havana in 1909–10 were about 2.5 times the average volume of the best three years between 1894 and

1899, Forbes suggested that increased nitrogen loading may have boosted fish yields through increased plankton production. In addition, higher water levels resulting in a larger acreage of submerged bottom lands would have been an extra benefit for feeding and breeding fishes.

Nevertheless, Forbes asserted that increasing reclamation of bottom lands into levee districts after 1907 probably began to counterbalance increased plankton production as backwater lakes, ponds, sloughs, and marshes were drained, and areas of overflow were reduced. And tellingly, Forbes and Richardson pointed out that in 1908 the largest yields of fish on the river were taken from exactly those areas with the largest acreage of connecting backwaters, where the plankton and bottom fauna (benthos) were most productive. Further reduction of these areas, Forbes predicted, would result in nutrients and plankton being sent more quickly downstream and out "into the Mississippi unconsumed" and leading to eventual lower fish yields. Forbes prediction was later confirmed when the increase of levee districts from 100,000 acres in 1908 to almost 150,000 acres in 1913 was accompanied by a precipitous 62 percent decline in the fish catch.

During 1909–10 Forbes sensibly divided the Illinois River into seven sections from its source at the Kankakee River (see fig. 5) to its mouth at Grafton on the Mississippi. Each section would now be studied in terms of its bottom, slope, tributaries, chemistry, sewage inputs, plankton, and fauna and flora. His objective was to investigate the river "as a unit with reference to its economic value, its protection, and its improvement." Robert Richardson—who had recently returned after two years in California working with David Starr Jordan at Stanford in ichthyology and ecology, and during 1906–8 with the U.S. Bureau of Fisheries—was to be in charge of field studies. In time he would be the principal author of papers generated by the work. Forbes was explicitly the overall director of the investigation, its conception, procedures, and scope.

Richardson's first task was to study fish breading areas. These were often in shallow water and were experiencing growing losses on the river. Richardson and his assistant, riverman Henry Allen, literally lived on the river during 1909–10, watching movements of fish and observing fish nests and young through a special viewing box mounted on the stern of their river skiff. Some former breeding areas were now too deep, and newer more shallow areas were not "seasoned" and not always attractive to fish for nesting.

As levee districts continued to enclose the floodplain, lakes and marshes shrank and the fishery shrank with them. A "fishermen's union" formed to oppose hunting clubs, and large commercial fishing companies bought extensive acreage on the floodplain and kept out small fishermen. Some hunting clubs turned wetlands into farms promising higher profits. Fishermen, fish dealers, and some farmers tried to delay levee construction, hoping for support from a Submerged and Shore Lands Legislative Investigating Committee (1909) looking into the encroachment of private persons and companies on the public waters of Illinois. The committees' report led to the formation of the River and Lakes Commission (1911), which heard citizens' complaints about lack of public access and the leveeing of bottomland waters. For awhile citizen access was protected and some water conservation successes were gained. But many a fisherman and hunter foresaw the end of their way of life and the destruction of public fish and wildlife resources on the floodplain, where for as long as they could remember the river rose and fell in cycle with the lush landscape and its plants and animals.[11]

In this atmosphere, Forbes and his colleagues resumed full-scale biological station operations. Of high priority were observations, biological-chemical collections, and measurements on the upper Illinois River in 1911–12, where Forbes expected poor conditions because of sewage diversion from Chicago. Indeed, graphically septic scenes greeted him and Richardson in midsummer from the river's source at the Kankakee downstream 42 km (26 mi.) to Marseilles (see fig. 5). Sludge banks oozed bubbling sewage gases like methane, carbon dioxide, and nitrogen to the water surface, and oxygen levels of water averaged only 7.5 to 9.8 percent of saturation. Characteristic pollution indicators or low oxygen organisms—including bacteria, fungi, rotifers, tubificid worms, and midge larvae—occurred abundantly in the malodorous water or bottom muds west to Ottawa in all seasons. Despite lengthy effort, they collected no fish in this segment of the river although active fish were observed at the mouth of the Kankakee and other feeder creeks in cleaner water.

Pollution indicator organisms dropped off quickly in abundance below Ottawa. Typical clean-water benthic animals—amphipods; crawfish; may, caddis, and dragonfly larvae; mussels, sponges, and Bryozoa (moss animals)—were increasingly abundant west of Ottawa and downstream to south of Chillicothe in upper Peoria Lake. Forbes believed that using specific benthic organisms like these as an "index to grades of contamination" was a promising way of showing the cumulative effect of previous environmental conditions in and on the bottom as had been proposed for European waters. Richardson took careful observations during this work and with Forbes would later (1921–28) put a biological pollution index with benthic organisms to extensive use for the first time in the United States.

During 1911–12, not until Hennepin, 106 km (66 mi.) from the river's source, did a clean-water component of fish species appear, and not until Henry-Chillicothe, 123–150 km (76–93 mi.) away, did typical relative abundances of these species appear. In a 1912 letter concerning a private suit brought against the Chicago Drainage Canal for the destruction of fishing, Forbes wrote: "It is certain, from the work thus far done, that part of the upper course of the Illinois has been rendered commonly unfit for the maintenance of fishes."[12]

During this second decade of the century, Stephen and Clara enjoyed the freedom and activities of a healthy couple in late adulthood. With his head professorship in entomology, a modest Army pension, and frugal habits, they spent their share of time away, always linked by effusive correspondence with each other or with family. Clara visited in Arizona, California, Indiana, Ohio, and Michigan. For his part Forbes admitted to her that he found it "easier to live with women more," between the times he spent with men at meetings, on trips, or in extended work. Consequently they found their 1912 European trip a joy, the success of which was dependent on his being able to leave the Illinois River work in confidence with Richardson and his field crew, supported in part by U.S. Bureau of Fisheries' funds. Back home when Clara was on the road, Stephen botanized, hunted quail, and caught up with nonscientific reading. Late in the fall of 1910 he had lectured again in Normal, then took a stroll and passed their old house from early marriage days. "It was newly painted," he wrote Clara, "but not changed at all. . . . What a flood of delightful memories!"

While her parents were in Europe, Ethel (Forbes) Scott, her husband, Frank (a professor of English at Illinois), young daughter, and friends vacationed on the eastern shore of Lake Michigan near Onekama. Exploring southward, they stumbled on an attractive piece of lakeshore north of Manistee and purchased it that fall. Professor Scott divided up the land, and sold a lot to Forbes, who built a cottage there next to the Scotts in 1914. A total of ten cottages went up by the early 1920s, all occupied by Illinois faculty families from English, entomology, history, law, and the library. There on the lovely sandy, wooded bluff above the lake, they named the site, "Dunewood." Over the next dozen years the Forbes spent time there—Clara far more than her husband. They needed a horse and wagon or a motor launch to get there. Light was by kerosene lantern; water was carried up from the lake. Later a cistern was built, filled by a gasoline engine, and the water gravity-fed to the cottages.

Richard Forbes, Stephen's grandson, who was there in the early years only as a small child, describes Dunewood as other family members remembered it:

> Life was very informal there, friendships were easily maintained, the lake was a center for relaxation with swimming, picnicking, canoeing, beach fires and beach walking. Tennis and horseshoes were available as were Saturday night dances, and costume parties, drawing, painting, playreading, and Saturday night "sings." It was a most congenial group. . . . Tom Scott [Ethel's son] used to say that he could not imagine life without Dunewood.[13]

Nor could many people imagine the Illinois River without its intact ancient floodplain, but year-by-year more of it was leveed. By 1913 bottomland in the lower valley in proposed levee districts boasted an average value of $30 an acre. Land already in the districts was valued almost four times higher, despite the fact that some levees had been unable to hold against the flood of 1913. The flow of the Chicago Drainage Canal alone (7,191 cu. ft./sec.) that year amounted to 86 percent of the flow of the Illinois at Peoria before 1900. "Flooding," of course, was exacerbated by restriction of the river to a narrower channel without the extensive natural absorption of the unleveed plain.

Forbes met that fall with officials of the Illinois Fish and Game Commission and the Conservation Commission at a special convention where he soberly read his summary on the river covering the salient facts. His greatest concern was the reduction of fish breeding and feeding grounds, especially in the lower river below Meredosia still far from serious pollution effects. How these grounds could be controlled and improved was the biggest challenge. Even in Europe this problem was "untouched" as he found on his recent trip, for example, on the Elbe where only some work was in progress. He had been to Washington as well and had talked with U.S. Fish Commissioner, Dr. Hugh Smith. The Illinois conditions were "largely new to their experience," Smith said, and a solution to them "has not been attempted in this country." Any federal government attempt to deal with it "would have to be an experimental project."

Soon after, Richardson observed a luxuriant growth of aquatic plants (*Potamogeton, Ceratophyllum, Scirpus,* and *Vallisineria*) with a rich benthic and weed invertebrate fauna in 1914 covering several square miles of Peoria Lake; in 1915 the lower river from below Peoria to Beardstown still contained a clean water benthic fauna. His and Forbes's extensive pioneering work on the benthic and shore fauna from Chillicothe 290 km (180 mi.) south to the river's mouth at Grafton during 1913–15 was based on 772 collections made with Blake and Ekman dredges, as well as collections from submerged aquatic plants (he called them weeds) in lakes and backwaters. The overall average weight of fauna living in or on the bottom in the main river channel from Chillicothe to Grafton he estimated at only 261 pounds per acre, whereas the standing crop in the more productive lake and backwater section from below Pekin to La Grange averaged 705 pounds per acre.

But even more significantly, within the shallow weedy areas of those labyrinthine backwaters and lakes, especially near Havana, also lived on and among the aquatic plants an estimated standing crop of 1,000 to 2,600 pounds per acre of small invertebrate animals. Combining both the bottom and weed animals from the river section stretching 30 km (19 mi.) above Havana showed that shallow backwater areas were at least twice as rich in potential fish foods as the adjacent main river channel. Again it was precisely those sections of the river having the most extensive connections to these shallow backwater areas that produced the heaviest fish yields. But how long would it last?

Some optimism reigned in 1915 when the River and Lakes Commission, with input from citizens and officials including Forbes, declared twenty-two lakes, one-quarter of the remaining unleveed ones, as public waters. Five of these bottomland waters were adjacent to Havana. It remained to be seen if the commission could protect the fisheries and other natural resource interests in the face of intense pressures for reclamation. Forbes said that he would continue his work on the river "in the face of the gigantic interests—agricultural, industrial, commercial, and political—which are now mustering along its course, with huge schemes in hand for revolutionary operations upon its channel, its banks, its tributaries, and its backwaters." In short, as always he would give it his best.[14]

TOWARD AN
ENVIRONMENTAL
SCIENCE

CHAPTER 9

Chief of the Survey, 1917–23

D uring 1914–15 the Illinois General Assembly scrutinized several bureaus and agencies concerning their economy and efficiency. A special committee, for example, found no clear line of distinction between investigative and scientific government functions and purely administrative and police functions. Reasons were readily apparent: agencies had appeared at different times and for different purposes, resulting in an awkward administrative structure allowing less than adequate coordination between them. Furthermore, several agencies had a variety of relationships with the University of Illinois that needed to be clarified.

The administrative code that emerged in 1917 under Governor Frank O. Lowden defined nine executive departments. The Department of Registration and Education (DRE) included the newly formed Illinois Natural History Survey (INHS) with Forbes at 73 as its first chief. The survey was formed by combining the former State Laboratory of Natural History and the State Entomologist's Office. The DRE also had responsibility for state normal schools and colleges as well as several licensing boards for trades and professions. In the DRE, an advisory and unpaid Board of Natural Resources and Conservation administered the INHS.

The immediate work and direction of the INHS after reorganization proceeded smoothly. However, the impending entry into World War I by the United States threatened labor and supply shortages and ultimately had a serious effect on survey work. Forbes was forced to cut programs with no vital economic interest. During the period 1917–23, from wartime through a postwar recession, INHS lost animal taxonomists and curators through transfers and deaths, and found it difficult to hire another economic entomologist. On a positive front, by 1922 INHS boasted four sections: aquatic biology, economic entomology, forestry, and plant pathology. The survey also maintained and increased a high level of cooperation with the geological and water surveys of DRE and with the University of Illinois. Forbes detailed these benefits for INHS to DRE director Francis W. Shepardson as: free quarters, heat, and water; the same use of university facilities as for faculty; unlimited access to appropriate graduate students; reciprocal use of faculty and researchers; and sharing of equipment and literature.

Forbes's major project as a pioneer in environmental science continued to be the Illinois River and the effect of pollution and reclamation on its aquatic biological resources—information "nowhere else available to those representing the public interest." Meanwhile, pollution crept past La Salle around the great bend of the river, and then southward during 1915–20 to do its most insidious damage. Forbes's and the survey's primary task was to document the resulting effects scientifically. He did not want to be pulled off his program and become embroiled in matters clearly outside the scope of responsibility and expertise of his survey scientists.

On the other hand, Forbes also argued for more rational state management of the fisheries using the knowledge in aquatic biology being generated by the INHS. Neither the Illinois State Fish Commission nor its successor, the Division of Game and Fish, was set up to do that, he wrote Director Shepardson. Furthermore, he reminded him that the comprehensive 1915 report of the Rivers and Lakes Commission on the Illinois River recommended that this state matter be resolved, so as not to allow strong agricultural pressures for reclamation compromise the integrity and survival of the fishery. Indeed, Forbes's contacts with the U.S. Fish Commission demonstrated to him that the federal people wanted state action before they contemplated cooperation. In a paper with Richardson about recent changes in the Illinois River, Forbes closed with this diplomatic warning:

> That the condition of the river and meandered lakes could be greatly improved by carefully considered management as a public fisheries property there can be little doubt, but the problem of methods of operation and policies to be adopted in such a case is a new one, for the solution of which there are no precedents. As a fisheries property the waters of the state are an unsubdued and neglected wilderness, and the investigator in this field must be prepared to do a pioneer work.[1]

With his survey work, responsibilities as the head of the entomology department, and the state entomologist's position he was about to shed, by mid-

1917 Forbes grew weary with his burden of work. He envied Richardson with his single-minded focus on their Illinois River project saying he was fortunate to have continuity in his work "day after day without the scores of interruptions and diversions which come to me here." So much for the benefits of office and fame. After he had just put new pistons in his car, he wrote Clara, "I shall certainly whirl out of town one of these days for a long run around the state," maybe even visit her at Dunewood for a few days. Forbes enjoyed being with people, but "lived so largely in a world of ideas," said his son Ernest, "that he did not feel the need for extensive social relations," even with relatives whom he preferred to stay in touch with by letter or phone. Still, in characteristic style he filled his days with volunteer activities: the board of managers and executive committee of the Champaign County Anti-Tuberculosis Health League, chair of the Champaign County chapter of the League to Enforce the Peace, and the Committee on the Preservation of Natural Conditions for Ecological Study, part of the Ecological Society of America.

Perhaps the proudest moment of his scientific career came in April 1918, when he was notified of his election to the National Academy of Sciences. A large and impressive reception was given in his honor on May 29 at the University of Illinois. Together, Clara and he were showered with accolades, kind greetings, and humbling praise.

Men and women are elected to the National Academy of Sciences (NAS) in recognition of their distinguished and continuing achievements in original research. It is one of the highest achievements a scientist can reach. Forbes was a founder of ecology in the United States and of economic ornithology, and a pioneer in aquatic biology, crustacean systematics, economic entomology, environmental science, insect pathology, and limnology. An ecological point of view dominated his entire research program before ecology was recognized as a distinct science. He firmly stood at the junction of pure and applied components of his research interests as he worked for an enlightened relationship of humans with nature.

Leland O. Howard, NAS member, economic and medical entomologist, and nominator of Forbes for the academy, drew special attention to the significance of all Forbes's early work emphasizing "his great breadth of view as a zoologist and entomologist. Forbes's work in entomology made a profound impression upon everyone interested in the applications of sciences to agriculture. . . . It will be difficult if not impossible to point out a naturalist of his generation who was more original or broader or sounder."

Forbes's elected class of fifteen included one person each from botany, chemistry, mathematics, neurology, paleontology, pathology, physiology, and zoology (Forbes); two each from physics and astronomy; and three from engineering. Forbes and the neurologist, Charles J. Herrick, joined the academy's section of zoology and animal morphology making a total of twenty-two section members. The NAS on December 31, 1918, had a total membership of 167. The NAS members closest to Forbes's research areas in 1918 who would have voted on his election included Charles Davenport, genetics and zoology; Leland Howard; C. Hart Merriam, mammalogy; Raymond Pearl, statistics and population growth; A. E. Verrill, invertebrate zoology; and William M. Wheeler, entomology.

More prosaically, later that year after more than three years as itinerant apartment occupants, the Forbes again moved into a house of their own at 706 Indiana in Urbana. Clara continued her work on the board of the Julia F. Burnham Hospital and as a trustee of the Unitarian Church. That fall Forbes was in Springfield as usual testifying before the general assembly about his biennial request for funds. He wrote Clara, who by now had recurring weakness of heart symptoms: "I think of you a hundred times a day, and always with the warmest love, tinged with just enough anxiety to make it all the more devoted."[2]

Come summer, Richardson repeated the sampling and analysis of the shore and underwater benthic fauna of the Illinois River. Forbes's plan was to do this again in 1920, and compare all the data with those obtained in 1913–15. Richardson worked the 165 km (102 mi.) stretch of the mid-to-lower river between Chillicothe and La Grange, halfway between Beardstown and Meredosia (see fig. 5). The situation was alarming. Pollution effects were immediately apparent in a previously less-affected part of the river.

By now citizens of the river valley were up in arms about Chicago sewage, adding, of course, to their own. A year later Richardson addressed the Peoria Rotary Club about the whole matter, laying out the status of the river, its value to citizens, and implications for the future from a biologist's point of view. An editorial in the *Peoria Journal* following his address reminded readers that the law creating the Chicago Drainage District was clear:

> All solid matter must be taken out of the Chicago sewage before it is allowed to enter the open river. The downstate districts should arise in righteous wrath and force the Chicago Drainage District to live up to the law. If the sewage kills the fish it is not helping humans. It would be much better to get back to the days when the odor of apple blossoms, clover blooms, and even sweet clover were enjoyed rather than to be forced to have our olfactories choked up with the smell of decomposing garbage and refuse. Shades of Marquette, Joliet, and La Salle!

Pressure had been building on the Drainage Commission, and finally in early November 1919 the Sanitary District of Chicago voted to construct sewage plants, the first at Maywood, an activated sludge facility on the Des Plaines River, directly west of downtown Chicago, and later a second plant at the stockyards.

To mend fences and share information the sanitary district proposed a meeting in late February 1920, inviting Richardson and saying that its officers would like to discuss pollution problems on the Illinois River. Forbes could not attend because of bronchitis, but urged Shepardson that Dr. Edward Bartow, chief of the water survey, should be there, since INHS did not have responsibility for protection of waters against pollution and Bartow's outfit did. The meeting "went well." Officials of principal towns from Joliet to Havana were present. Bartow, a straight-shooting man, said that Chicago was responsible for the pollution, and that the Drainage Commission was trying to provide a remedy which might take "several years" to become effective. Forbes, in a letter to Shepardson, thought that the commission should be "let alone and be allowed to do their job." To all officials who would listen Forbes emphasized that sewage must be treated—dilution was not enough—and particularly that the level of meat-packing and other manufacturing wastes was far above current state laws and must be managed.

For ease of work and better mobility Richardson now had a gasoline launch for his benthic work loaned from the State Division of Game and Fish. Forbes occasionally went along but spent much of his time spreading the word and asking for professional state and federal help for the river work. This paid off: the U.S. Public Health Service embarked on an exhaustive study of pollution conditions with headquarters at Peoria. Forbes was one of three consultants on the project appointed by the surgeon general of the United States. And then there were his many trips to Springfield concerning his funding. Forbes wrote to Clara:

> You do not know, you can hardly imagine what it is for me to be away from you for so many days. . . . When I think of you it is the real, essential you that I see—the unchanged, unchangeable one whose picture you sent me from La Porte [in the early 1870s]. That is the wife I live with when I am alone. . . . This, I suppose, is the way we take with all the objects of our worship, and how we are lifted above our ordinary selves, as I was this morning. I wish you had been with me. Goodnight, my dear one, A.[3]

Forbes odyssey through science in the public service was now at its zenith. His reputation and influential capital were available at his own choosing. In 1921 Forbes retired from the University of Illinois faculty, and now for the first time in his almost fifty-year professional career, he wore only one hat— chief of the INHS. No doubt remembering his 1916 talk at New York meetings of the Ecological Society of America (ESA) on the salient connections between ecology and economic entomology, and how they could profitably work together, the society recognized an opportunity and elected him president for 1921. In Forbes the ESA gained as president a man who saw ecological theory and pure science as complementary to practical applications and for the benefit of society.

Forbes's presidential address that December in Toronto considered the proposition that the field of ecology should be brought into much closer relations with human enterprises and welfare. Forbes felt that because ecology cuts across, joins, and affects other biological sciences, and given that humans shared the environment with other living organisms, "ecology is, in short, of all the biological sciences, the humanistic science *par excellence.*" To those who insisted that applied ecology (e.g., economic entomology, economic zoology, or bacteriology) was not part of ecological science, Forbes said this could "be justified only by setting man outside the general order of living nature in a class by himself." Hence "the ecological system of the existing twentieth-century world must include . . . man as its dominant species." In the spirit of this argument Forbes, as the leading American standard bearer for ecological work before it was even named as such, embraced views similar to those held by many other early ecologists who were involved with practical aspects of agriculture, forestry, and horticulture. Then, as now, the pure and applied proponents of ecology had much to learn from each other—and from Forbes, who bridged them both, and who now offered a view toward a truly environmental science.[4]

In 1921 Richardson published the results of their most recent work on the Illinois benthos. Between Chillicothe south to La Grange, he estimated a 34.5 million pound reduction in summer benthic animals from 1915 to 1920. This amounted to an average 75 percent reduction in the earlier "clean water" fauna and a commensurate increase in sludge tubificid worms and midge larvae, with polluted conditions advancing southward from Chillicothe to Havana by about 20 km (12 mi.) per year. He measured dissolved oxygen at the water's surface as low as 1 ppm at Havana during midsummer 1920. For the entire middle river, Richardson predicted a potential drop of about 7 million pounds of fish yield from a river that was only a short time ago "among the richest of its kind in the fresh-water stream systems of the world." By 1920 fish were much less abundant as far south as Peoria Lake, and the lush growth of submerged aquatic plants that Richardson had seen there in 1914 had largely disappeared. Since 1910 Chicago's human population had increased by almost 25 percent to 2,701,000 and the stockyard sewage alone—although the industry had just come through a recession—was additionally equivalent to that from a city of one million people.

Richardson continued his fieldwork during 1922–23, primarily on the middle portion of the river from Peoria to Beardstown. Much to his and Forbes's satisfaction, sewage plants had by now come on line materially improving the Chicago effluent; nevertheless, more than two decades of pollution's insult had taken a toll reflected by the river's benthic communities. All of the benthic species north of Peoria they now categorized as either "pollutional," or as a more or less "pollution-tolerant" group of tubificid worms, leeches, midge larvae, and fingernail clams *(Musculium* and *Pisidium)*. Several dozen more sensitive clean water species (including mussels) present in 1915 were now not seen. Just north of the city in upper Peoria Lake, profuse concentrations of tubificid worms and fingernail clams reached 50,000 and 84,000 per square meter, respectively. South of Peoria 115 km (71 mi.) to Beardstown, benthic habitats failed to yield an average of three dozen (range of 13–69) clean water species that were present in 1915. After 1923, Richardson detected a shift to an increased presence of clean water species from Havana to Beardstown as conditions improved.

Much as he had predicted from the 1920 benthic data, the fish yield from the river for 1921 amounted to an all-time low of four million pounds, but then rebounded to 10 million pounds in 1922, yet still 25 percent less than the yield in 1913. Ninety percent of this 1922 fish yield came from the river below Pekin south to Grafton. From 1924 to 1930 the fish yield "varied around 10 million pounds per year."[5]

Concluding fieldwork in 1924, Richardson presented an analysis and summary of the Illinois River story since 1913. An important feature was a pollution index based on the benthic fauna, an index that he and Forbes had come up with and continued to modify during their twelve years of work. Most straightforward was the 1925 version. There were three groups: 1. Pollutional or more or less tolerant species—with tubificid worms, leeches, midge larvae, and tolerant fingernail clams; 2. Cleaner preference species—animals preferring cleaner water, but present in faster-moving water with some pollution; and 3. Typical clean water species, including most prosobranch (oxygen requiring) snails.

Richardson then presented an expanded but provisional list of indicator species in 1928. This had the following additions: former group 1 spun off three new groups based primarily on tolerances of different species of midge larvae. Then he divided former group 2 into three new groups as well, including some small clams, less tolerant leeches and worms, and air breathing (pulmonate) snails. Former group 3 was essentially left intact as new group 7.

Most important, Richardson, with Forbes's explicit approval, cautioned against overconfidence with lists of indicator species and the laying out of "hard and fast lines" of pollution zones especially by inexperienced biologists. They had learned, for example, that different species of midge larvae and some species of sphaeriid clams exhibited a much wider range of pollution tolerance than previously known and that some accepted "pollution species" were frequently found abundant in clean water. But even more basic, the abundance and distribution of many potential pollution indicator species were often so variable, even with diligent sampling, that for a particular river the biologist "is bound sooner or later to be forced back upon his own resources." This would include his accumulated knowledge as to the *groupings* of species under polluted conditions, compared to those observed under earlier clean water conditions. In short, kinds could be more informative than just numbers. And he urged the use of chemical data—as he and Forbes had done—for supplementing biological collections in assessing the effects of pollution on the benthos. The accumulated twenty articles of almost 1,900 pages by Forbes and his colleagues on the Illinois River during these years had a deep and far-reaching effect on the practice of aquatic biology both here and abroad.

By 1923, with the exception of the benthic work, Forbes felt the Illinois River project now concerned primarily sanitary matters more in line with the water survey and its role in maintaining the quality of public water supplies. After receiving formal approval Forbes turned over the INHS floating laboratory to the water survey's new chief, Arthur M. Buswell. Forbes then proposed a new project in aquatic biology: "I would like to take up a comprehensive, systematic survey of the waters of the state, one river system after another, to be studied as features of our natural resources, especially for recreation, scientific study, and the production of food."

Accordingly, Forbes transferred his main operations to the Rock River system in northern Illinois, including its main tributaries the Pecatonica and Kishwaukee, making up a watershed of 28,000 sq. km (10,810 sq. mi.). His center of operations here lay at Rock Falls midway on the river between Rockford and Rock Island on the Mississippi, with Dr. David H. Thompson in charge. Thompson and four men, with their "navy" of a cabin boat, a launch, and two skiffs, spent three years on biological and chemical studies of these rivers, other small streams, sloughs, and ponds. Fish were also tagged to determine their growth and movements. Meanwhile, Richardson and his own crew of 2–4 men in Urbana received and processed the accumulated material and data. Forbes ran these operations on $10,000 per year. By the late 1920s Forbes had moved on to similar work on the Sangamon and Kaskaskia Rivers and on smaller streams in Champaign County.

Information generated by all this aquatic work led to additional cooperative projects with the Illinois Department of Conservation involving fish culture, habitat work, and stream stocking—activities that Forbes had long believed should materially benefit from his INHS initiatives. Furthermore, as a member of the National Academy of Sciences, he had recently (1925) been appointed to the Committee on Aquaculture of the National Research Council, whose charge was to examine the status and needs of aquacultural enterprises in the United States. Forbes believed that beyond the necessary scientific work required, an important need existed for bringing the appropriate information generated to the people doing the practical work, whether private, state, or federal. Dry land farmers had their experimental stations and county farm advisors; aquaculture had neither.[6]

Ever since his painstaking work on the food of birds during 1877–82, Forbes had wanted adequate relative abundance and density data on bird species over a variety of habitats to determine their potential ecological role and their economic effect on agriculture and horticulture. It occurred to him that an equivalent of the plankton method might work, that instead of a net, two men would walk in a parallel line, say 20 or 50 meters apart depending on habitat, over a definite distance, identifying (shape, size, color, flight, song) and counting "all the birds flushed by them or crossing their track," making at the same time notes of the habitats. Forests of tall trees would not be covered, nor would aquatic birds be counted. This unique sampling method was first tested by one of Forbes's undergraduate students, Alfred O. Gross, and a friend on the university's 400-acre grain and stock farm. Then a full year was spent in this fashion in three sections of the state—north, central, and south—during each season. Additional trips were made in the summers of 1908 and 1909. The press of other duties, and Gross pursuing a doctoral degree at Harvard, precluded completion of this work. It was not until 1921–23, when Gross was a professor in Maine, that the two men finished and finally published three papers on all of the work carried out on foot over a total distance of 4,520 km (2,802 mi.) when close to 65,000 birds were identified. Unlike the typical bird census at the time, theirs produced for the first time more precise abundance and density data over various habitat types with greater scientific value compared with the more vague descriptions of "abundant" and "scarce" usually offered for bird species.

Their summer work yielded 133 species of birds encountered while tramping through more than 19,000 acres of Illinois meadows, pastures, swamplands, bare fields, grain (5 kinds), stubble, plowed ground, woods, orchards, yards and gardens, and shrubland. Eleven species of birds made up three-quarters of the total numbers (22,500) of birds seen. Five of these eleven species—English sparrow, blackbird, meadowlark, cowbird, and mourning dove—were most abundant (1,131–5,653), were most prominent in terms of numbers per square km (14–72), and were represented among the principal birds for each section of the state. They also occurred in nearly all habitats listed above, absent only a few times in shrubbery, swamp, and woodland.

Of the fifteen habitat types, the orchard was decidedly the favorite for birds on a statewide basis with 52 species present. Although the overall average summer density for birds in the state was 329 per sq. km, the density for only 118 acres of smaller farm orchards was 1,554, nearly five times as concentrated. Following orchards in order of density were yards and gardens (1,202), swamplands (704), and woodlands (568). It bears noticing here that birds of human neighborhoods recorded twice the density of birds from more wild swamp and woodland.

Winter contrasted strongly with summer as expected. More generally in flocks, birds swept "widely from place to place as the weather changed, in a free search for shelter and a comparatively scanty food supply." On a trek of 587 km (364 mi.) through more than 6,000 acres, they encountered 5,193 winter birds of 52 species averaging 201 per sq. km, or 61 percent the density of summer birds on a statewide basis; in contrast, there was an almost 10 percent higher density in southern Illinois compared with summer because of northern birds migrating south. Unlike summer the top three most abundant species were not overly dominant in numbers. Statewide, the crow, lapland longspur, and junco ranked 1–3 in both abundance and density (567–768, and 10–30 per sq. km).

By and large, winter birds were primarily seed eaters with largest numbers in cornfields, pastures, and woodlands. Although densities of birds were quite large in orchards, shrubs, and yards and gardens, the very small acreage involved (127, or 2 percent of total) and extreme gregariousness of the birds made these statistics less reliable. Winter densities of the top two species in each of the three favored habitats were: cornfields—crow (71 per sq. km) and junco (24); pastures—goldfinch (52) and prairie horned lark (37); and woodlands—junco (112) and blue jay (44).

During spring migration all was in a state of flux, with birds in a "swirling stream," but pastures and meadows were most densely populated (421 and 419 per sq. km, respectively). When birds readied for fall migration, summer ties were broken and again pastures surpassed all other open field habitats in bird density (575). At the same time habitats with trees, shrubs, gardens, fallow, and scrub, as a group totaled 181 acres and boasted high bird densities averaging 1,335 birds per sq. km.

In the end they had statistically examined 166 bird species, and with an additional 29 species seen outside the surveyed strip, their total state list amounted to 195. Of the 166 examined species, Forbes and Gross estimated that 50–75 percent were abundant enough in a given sector of the state or in a particular habitat to give them a potentially significant ecological role and economic effect on agriculture and horticulture.

With an eye on the future, even at age 79, Forbes showed empathy for the reader in this paper and for bird lovers in particular. "Probably nothing can seem further removed from the source of interest in ornithology . . . than a mass of statistical data." But with imagination, he thought, one could translate the figures "into a captivating vision" of the actual bird life of an area, "to make the picture as true as possible to the life of field and forest, thicket and swamp, summer and winter, north and south." Forbes and Gross had presented the broad picture, the background tone and tint. They planned a concluding paper on final species detail "all over the state and around the year," but it was not realized.[7]

Given his sustained interest in state bird habitats, it is not surprising that Forbes had a leading role in forest management and conservation as well. Back in late 1907, Thomas Burrill had advocated a thorough forest survey of Illinois in a paper at the annual State Horticultural Society meeting. He reminded members that the State Laboratory of Natural History under existing law had the power to do this. Forbes assented. The society approved the idea, appealed to the U.S. Forest Service for assistance, and prepared for discussions with state officials regarding a needed state forest policy. With the next legislative session several years away, Burrill decided an immediate commercial survey of existing forests should be made. Sharing both management and expense, the state laboratory and federal Forest Service made the survey in 1908–9. The survey covered 1 million acres in 26 counties (80 percent in southern Illinois), making up about 30 percent of the total forest acreage in the state. Most forests were in poor condition, had low productivity, and lacked proper management; consequently, the forest products industry lay in decline. As a result, a forest policy bill was prepared and presented to the legislature in 1911, but it never received a hearing and died in committee, much to Burrill and Forbes's disappointment.

It will be recalled that Governor Frank Lowden's 1917 administrative code provided for forestry matters concerning investigation and scientific work with the DRE and its Board of Natural Resources and Conservation. On the other hand, the promotion of forestry as an agricultural industry lay under the aegis of the State Department of Agriculture. So Forbes recommended to the board that a forester be appointed to his INHS staff. Forbes was especially pleased with Lowden, who had personally engaged in large-scale forestry work on his own spacious Sinnissippi Farms, and who in his 1917 inaugural address called attention to the "many hundreds of thousands of acres of land in this state better suited to forestry than to anything else. Private owners of land, however, will not content themselves with a crop which does not mature for half a century. They will, therefore, naturally not plant these acres to trees unless they have encouragement from the state."

During 1918 Forbes addressed the Chicago Academy of Sciences about both river and forestry problems. Lowden read reports of this paper and was "stirred a good deal" by it. Wanting "to keep the word forestry before the people," he planned to include a paragraph on forestry in his next biennial message to the state.

Without a forester yet on board, Forbes went ahead with plans for a rational forestry plan to present the governor. And he boasted about "a wholehearted group of Illinois naturalists," including the eminent University of Chicago botanist Henry C. Cowles, who had volunteered a study of tree growth in eight counties around the state in 1919 and asked only travel funds from Forbes to do it. That summer the Canadian forester R. B. Miller reported for duty at the INHS and immediately began preliminary studies and an educational campaign throughout the state.

Before the nineteenth century probably 30 to 40 percent of Illinois was forested. By 1920 nearly two-thirds of this primitive forest was gone, mostly to provide crop land that was soon worn out, eroded, and then abandoned or used for pasture. These mostly hilly lands made up approximately 15 percent of the state's area, some six million acres lying along the Mississippi and Illinois Rivers and elsewhere in seven southern counties—none suitable for farming, but some of it suited for growing trees. Forbes found it disgraceful that a prominent agricultural state like Illinois should have millions of acres of soil being so "mismanaged and misapplied." "So wasteful a policy should be corrected," and these soils "put to their best possible service." He was gratified that he now had the Illinois Farmers Institute's support for his developing plan, given that 90 percent of Illinois forest was in farmers' woodland.

With funds appropriated by the general assembly, Forbes moved ahead in 1921 with a systematic survey of forested and deforested lands in 78 counties. Three additional foresters were hired for the work, with C. J. Telford in charge. They proceeded to identify the location, composition, growth rate on various soils, and management of the existing forests, as well as estimating the potential of deforested sites for reforestation. With these data, Herman C. Chapman, professor of forest management at Yale University School of

Forestry, then estimated the value of Illinois woodlands, the economics of forest management and production, and the supply and demand of forest products in Illinois—all necessary for framing a workable forest policy.

Yet there was more to forests than just growing trees for a profit. Forbes expected a truly public survey to address other values and needs as well. Forests "are often of greater use for educational, recreational, and esthetic purposes, as parks," or as "preserves of the primitive life of the state, of great interest and value to the student of science and his teacher and lovers of wild life."[8]

Chapman's elaborate report at the close of the survey in mid-1923 identified five million acres of Illinois land "which are or should be in forests." This total included some two million acres of treeless, eroded lands that could be reclaimed and reforested. Chapman estimated that the entire state was capable of an annual production of 435 million cubic feet of wood, almost four times current production. Even farmers' woodlands—the largest portion of existing forests—were producing only half their potential.

With this report as an impetus, a bill was introduced in the state legislature and passed in 1923 for the establishment of a forestry division in the university experiment station. Financed by both the INHS and the U.S. Forest Service, investigative and extension work in forestry then went forward during 1925–30.

Until 1926 Illinois had no state forests as such, but had authorized instead the establishment of forest reserves by counties. Three counties had done so, with Cook County's being the most prominent. By an act in 1926 a newly created Illinois Conservation Department was authorized to organize state forests, designate portions of these as fish and game sanctuaries or public parks, market products from these forests, and provide forest tree nurseries for reforestation.

At an age when most people would have brought down their game a notch, Forbes had eagerly caught Burrill's pass back in 1907 and ran with it. Despite early failure, with his tenacity and leadership Illinois finally had a program for scientific and extension forestry. Once again Forbes had linked his name with science in the public service and the conservation of natural resources.

Nevertheless, during the mid-nineteenth to the first third of the twentieth century, the state of Illinois lacked a formal comprehensive policy for natural resource sciences or any other science. Rather, the state began and continued scientific activity primarily in response to specific needs reiterated through the requests of organized interest groups. Early group leaders included Cyrus Thomas and John Wesley Powell. In this fashion, state support of science grew gradually, sparked and sustained by strong groups and their politically astute, outstanding scientists and administrators, especially those leaders whose scientific work had wide distribution and potential economic effect across the state. Forbes and Edward Bartow, chief of the Illinois Water Survey, are later examples. Strong administrators at the University of Illinois, such as President Edmund James and Dean of Agriculture Eugene Davenport, as well as Governor Lowden, also took their lobbying efforts directly to the public, Illinois alumni, or business interests with substantial success.

Forbes's repeated theme was that applied and pure science in the state and at the University of Illinois had the obligation to "stand in the closest possible relation to the general public welfare." As we have seen Forbes worked tirelessly through vigorous publications and public relations to show the state legislature and citizens the importance of science for the state's economy and for the enlightenment of Illinois citizens. The state legislature responded in kind to Forbes's well organized and energetically delivered requests for support and grew to more fully appreciate his blend of applied and basic research in ecology. On the whole, Forbes was remarkably successful in these efforts, and he set a standard of leadership for scientific agencies in Illinois that continues to impress even today in the twenty-first century.[9]

Acknowledging his splendid success in a wide variety of scientific enterprises, one yet wonders why more zoologists in the United States did not pick up on Forbes's ecological lead. Certainly the majority of his early papers (1876–90) were reasonably available in this country and were also known in Europe during the last two decades of the nineteenth century. To be sure, a preponderance of these papers were entomological (77 percent), and by the late 1880s and early 1890s he was making a distinctive mark with economic entomological colleagues. Yet he also made a strong impression on other nonentomological colleagues such as David Starr Jordan, Spencer Baird, B. W. Evermann, Charles Whitman, A. E. Verrill, S. I. Smith, Jacob Reighard, E. A. Birge, and the European scientists F. A. Forel and A. Thienemann—all of whom were familiar with his aquatic biology work, and for the last three men just mentioned, especially with his microcosm concept, which they explicitly accepted.

There may also have been a lack of a critical mass of other ecologically oriented zoological colleagues who carefully read his early papers and understood the significance of his findings as impetus for their own work. Forbes was far ahead of his time in his thinking about, description, and application of ecological concepts and principles. Furthermore, some zoologists also may not have found his study material and methods particularly attractive, or they found the work required more time, patience, help, and funds than they had available. Those years were especially lean times for many, and few may have had his ability to design and generate funding for the type of integrated practical and more pure scientific program he would sustain for many years through changes in the economic climate. We are left with the certainty that Forbes was not the first pioneering scientist to travel a little-used road. With his breadth of vision, uncanny sense of direction, persistence, and leadership, he was a giant of American ecology.

CHAPTER 10

The Path before Him, 1923–30

C hristmas 1923 was the occasion of Stephen and Clara's fiftieth wedding anniversary. Forbes used the occasion for reflection and writing about their lives together and the meaning of it for their extended family when they gathered at the homestead:

One of the conclusions to be drawn from our experience is that of two persons as closely associated as husband and wife, one may sometimes be irresistible and the other may sometimes be immovable, but that they must enjoy these privileges turn and turn about, and that neither must ever be both at the same time. Take your agreements seriously, and your disagreements humorously, even if only one of you can be humorously inclined. Next to affection, humor is the best peacemaker. Do not hope to be or seek to become alike. Unlikeness, amounting soon to contrast, makes the life together interesting.

We know that whatever happens to us during the second half century of our married life, now just beginning, and however different it may be from the first half century just finished, as long as we live we shall always rejoice in the reflection that we have had as happy a life as often falls to the lot of man and woman—a half century of unbroken and unalloyed affection for each other, which has figured so largely in our experience that we can hardly imagine what life would have been without it, and this group of beloved children and children's children, who could never have lived at all if we had not first loved each other; and with these the daughter and sons of other parents, who have been drawn into our group as the lovers of daughters and son of our own.

Forbes then remarked about one thing he would have done differently—in the best of worlds. He would have finished a four-year college course, working himself through as necessary. And in so doing, "I should have come to know better what I was really most capable of and should have avoided any false start, such as that in medicine." But, he reflected, his life course had been set differently via teaching and his marriage to Clara. And *that*, in life's curious way, had now made all the difference in the world.

Of his inner life, about which he was usually reticent in person, Forbes wrote in 1923: "I was and still am, a rationalist and an agnostic, for whom what is known as faith is merely assumption, often practically necessary, since in active life one must very often act *as if* he believed what he does not and cannot really know, but inexcusable in purely theoretical matters."

However, Ernest Forbes, by then director of the Institute of Animal Nutrition at Penn State University, said in 1930 that his father drew a line through the words "and still am," and wrote above them, "as a younger man." Ernest went on: "I think that this change was dictated by a growing hope that there is more to life than a scientist knows, or can know. At this time he was much attracted by the beauty and comfort of the orthodox beliefs of his childhood, and which he had lost during his scientific career."

Forbes description of himself as a rationalist was an honest portrayal, one that further informs us of his view of himself, his work, and his role in human society as described by one of his favorite philosophers, Marcus Aurelius: "If mind is common to all, then also the reason, whereby we are reasoning beings, is common. If this be so, then also the reason which enjoins what is to be done or left undone is common. . . . If this be so the Universe is a kind of Commonwealth."[1]

As he approached 80, Forbes still drove his car with some regularity. Family anecdotes about Forbes replayed his minor mishaps and rapid driving. Even his grandchildren gloried in his racing side-by-side with the orange interurban railway train.

On his eightieth birthday in May 1924, Forbes got a late start to attend a special luncheon in his honor at the Rotary Club. After being stopped by a policeman for speeding, he was let go because of the impending event. That summer he wrote a friend that he planned "to spend the greater part of the summer in racing about the state on my business and [Clara's] pleasure. . . . I have at last taught her to permit me to run at an average of forty miles an hour, but never as much as fifty." That did not last. Three years later he drolly advised her to travel "by steamer, rail, or aeroplane," if she must, and not with him if his speed scared her.

In his eighties Forbes retained his usual close devotion to duty. He seemed to live mostly in the present and had the rare ability to stay connected to attributes of his fresh youthful self. University colleagues and people in general considered him a good neighbor and loyal friend. Among the INHS staff, they touted him for his honesty, sound judgment, and inspiring influence, and for his astounding familiarity and study in so many fields of knowledge. Younger colleagues found him a fair critic and praising of their efforts. In his later years he took keen interest in his financial and investment affairs but had little enthusiasm during his entire life for pursuing private financial gain for its own sake. By 1926 his yearly income from all sources amounted to about $5,800, and soon after he adjusted his investments in wary anticipation of an approaching depression, sacrificing some income for safety.

His chief attention, however, went to his science. To him science "was life itself," said Ernest. In the thirteen years since he had assumed his position as chief of INHS, Forbes published a remarkable twenty-seven scientific papers on entomology, aquatic biology, forestry, birds, and general ecology, maintaining his productivity to the very end.[2]

During the latter half of 1926 and on into 1927, both Stephen and Clara fell ill—he with a bladder infection, and she with lung congestion, but both recovered well. Following the illness Forbes prepared a paper for a National Academy of Sciences meeting that fall and administered the new state forestry program, and Clara traveled that August for the last time to Dunewood. At home they had a full-time housekeeper-cook, and friends or family who occasionally looked in. Forbes retained his customs, still cherishing a bowl of Grimes apples for munching or chocolates after meals, and he was always ready to share some of his favorite home-brewed beer.

Forbes was delighted to hear from David Starr Jordan in August 1927 who wrote: "We are both old veterans now, among the last leaves on the tree, but still awake." But Forbes also wrote Clara in the same month that despite his having "a good night's sleep and a morning resolve to plunge into the work as if my fortune was still to make," he left the office in the afternoon feeling either "too old or merely too tired for continuous hard work."[3]

Nevertheless, at 82 he still actively led the Illinois Natural History Survey. For example, by 1926 the voracious European corn borer, *Pyrausta nubialis* (Hübner), had reached Indiana and was soon expected in Illinois. In infested areas, corn yields fell 50 to 90 percent. As a biological preventative measure, Forbes had advocated use of the wasp parasite *Trichogramma minutum* of corn borer eggs. By 1928 his field entomologists were releasing 10,000–25,000 wasps every day. Eventually a joint state and federal effort plus droughty conditions reduced borer populations and their migrations, so that when the pest finally reached Illinois in 1939, INHS efforts had substantially helped to reduce its effect.

Then too, after a passage of two decades, Forbes resurrected his active work in public school science education, proposing in 1928 free distribution of selected sets of INHS publications for high school biology teachers and their students. First, two INHS biologists who had taught high school biology examined 200 publications, and chose 68 of the most appropriate. Second, these were examined by nine current biology teachers who shortened the list to 63. Almost two-thirds of these 63 were written by six men: Forbes (24), entomologist W. P. Flint (10), R. E. Richardson (3), the foresters C. J. Telford (4) and R. B. Miller (4), and David H. Thompson (2). Forbes sent sets of these publications to 142 selected high schools in Chicago and downstate from among those schools accredited for student admission to the University of Illinois. The fact that Forbes so late in life assisted the high schools in this way brought him wide esteem and much personal satisfaction.

By 1928 Forbes had served the state of Illinois for sixty years, and at 84 was its oldest employee and last Civil War veteran on campus. "In his uninterrupted service to the state," wrote an anonymous chronicler, "he held himself and his work rigidly aloof from political influences and commanded the respect of all parties." Rallying to his side in his last years, the Urbana Association of Commerce praised him for his "courage, devotion to duty, and high patriotism" in the Civil War, and for his impressing "upon all, the true principles of leadership, of truth, and service to mankind." Not to be outdone, David Kinley, president of the University of Illinois, suggested that a set of memorial volumes of Forbes's "pioneer papers" be brought together, and he appointed a committee to do this with Forbes's help, but the task was not completed at that time. Forbes had chosen thirty-four of his papers for a memorial volume, and it was not until 1977 that a volume was published, including nine of the pioneer papers suggested by Forbes with Frank Egerton, the ecological historian as advisory editor for Arno Press.[4]

Forbes's wife's health deteriorated again in 1928, and their daughter Ethel was quickly summoned for fear Clara would not recover. Ruth Kelso, a friend, stopped in to see her and Stephen asked Mrs. Kelso to speak softly so Clara would not hear and insist on visiting since he wanted her to rest.

"Who is there?" Clara called. And again but louder, "Who is there?"

"Ruth Kelso," said Stephen.

"I want to see her."

So Mrs. Kelso went in and found Clara sitting up in bed.

"Well," Clara said, "they thought I was gone this time. But here I am."

Mrs. Kelso remarked later that "she looked downright pretty in her pink wool jacket with a little air of triumph and a sparkle in her eyes. An indomitable soul she had. . . . More than anyone I have ever known she . . . took the whole community to her heart and made it a better place to be either through personal service to individuals or support for all good movements."

Clara improved somewhat but Stephen told family members that she would "never again be as physically active" as she had been in 1927. As 1929 wore on Clara occasionally displayed a "wandering mind." Stephen, after battling some throat infection and tiredness, decided again to leave his troubles "to the well-known *viz medicatrix naturae*."

By Christmas he sat with Clara almost constantly, as did her nurse. Besides, Stephen had been supplied with a heart stimulant by Clara's physician to be given to her in an emergency, and he lay awake frequently at night "full of foreboding" for her.

Soon after New Year's, Clara began losing ground as her heart progressively weakened, and she lapsed in and out of consciousness. She died in her sleep, January 24, 1930, at the age of 82, with Stephen at her side.

"She was the best woman I ever knew. We must draw close together," he told his family, "now that the strongest connecting line is broken."[5]

Three weeks later Stephen had grieved some, recovered from a cold, and moved around more confidently. He weighed in at 130 pounds and registered an equivalent systolic blood pressure. With daughter Ethel's help he pitched into a general cleanup and freshening of his house, with the addition of a new rug and reupholstered chair. When she left for home—at his insistence—he sat down and drew up plans for rehabilitating Clara's old-fashioned backyard garden, which had gone wild. Late in February he contracted with two live-in people completely relieving him of chores so he could finish up his annual Illinois Natural History Survey report and put final touches on a few papers. Early in March he had several chills and told his doctor that he was "too busy to be sick." But by the ninth of March he had faded considerably.

Forbes was admitted to McKinley University Hospital where he died at 6:30 A.M., March 13, of "inflammation of the bladder and complications," three and a half months short of his eighty-sixth birthday. Bertha and Ernest were at their father's side. Ernest portrayed his father's passing best: "The final, supreme demonstration of his superb morale was when with health, strength, friends of his own age, and wife all gone, he was able, through sheer intellectuality, to continue at his work, stimulated by the doing of new things, in the true spirit of youth."

Among the papers found after his death were the following lines in his writing, and with his initials:

He is not old who loves the young,
Whom the young love is young himself.
The full heart is the happy one.
May the full heart–the curious mind,
Be yours until your latest day.

Then shall your age be fresh as youth,
And late December bloom like May.[6]

Forbes's funeral service on Sunday, March 16, in Smith Memorial Music Hall on the Illinois campus drew hundreds of people. The occasion expressed the tapestry of his life: part solemnity, part celebration, part thanksgiving, part fraternity, part comradery, and part family.

Thirty-five honorary pallbearers included representatives of the National Academy of Sciences, the Illinois Academy of Science, the University of Illinois faculty and administration, and other state bodies, and his eminent scientific colleagues and friends, including Henry C. Cowles, Stanley Coulter, Edward A. Birge, David Starr Jordan, and Leland O. Howard.

The detailed eulogy of the day was neatly summed up soon after by his colleague and successor as head of entomology, Clell L. Metcalf, who recited Forbes's qualities:

> Fearless, eternally youthful, unostentatiously confident and inspiring, never seeking favor or preferment but continually in demand by recognition of his worth, this man was revered by his peers and colleagues for his breadth and clarity of vision, his kindly, helpful criticism and sympathy, his infectious enthusiasm, his brilliant intellect and impregnable strength of character, and his loyal and genial friendship.[7]

And for the defining core of the man made manifest from his wartime cavalry days, Rev. Edward Johnson, Pastor of the Urbana Unitarian Church, spoke the fitting lines of William Wordsworth:

> Who is the happy Warrior? Who is he
> That every man in arms should wish to be?
> —It is the generous Spirit, who when brought
> Among the tasks of real life, hath wrought
> Upon the plan that pleased his boyish thought;
> Whose high endeavors are an inward light
> That makes the path before him always bright;
> Who, with a natural instinct to discern
> What knowledge can perform, is diligent to learn.[8]

Soldier, teacher, scientist, administrator, guardian of the public welfare, health, and enlightenment, and champion of the environment, Stephen Forbes saw life and nature whole. Looking forward with a full and generous heart, he persevered to the last, with duty, service, and science.

Appendix 1. A Soldier's Library

The following books were used by Stephen Forbes during his Civil War service, 1861–65. (Source: Scott, Ethel Forbes [1936], "Forbes Family Letters and Journals of Stephen Alfred Forbes, 1847–1870," p. 520 and addendum.)

Addison's Works. Vol. 4. Miscellaneous Prose. 1860.
The Book of Common Prayer. Philadelphia: King and Baird.
British Essayists. Vol. 7. Alex Chalmers, ed. London: Johnson, 1802–3.
Bullion's Greek Grammar. New York: Pratt, Woodford, 1850.
Cavalry Tactics. Parts 1 and 2. U.S. War Department. Washington, D.C.: Government
    Printing Office, 1863.
Chefs-d'Oeuvres de Jean Racine. New York: Ivison and Phinney, 1858.
A Grammar of the Greek Language. Alpheus Crosby. Boston: Crosby, Nichols, 1859.
Greek Lessons. Alpheus Crosby. Boston: Phillips, Sampson, 1858.
A Greek Reader. Charles Anthon. New York: Harper and Bros., 1842.
Green Mountain Annals: A Tale of Truth. Garret Van Hoesen Forbes. New York: Burnett
    and Smith, 1832.
The Holy Bible. London: T. Nelson and Sons, 1854.
Julius Caesar's Commentaries on the Gallic War. With notes by E. A. Andrews. Boston, 1860.
A Key to the Exercises in Vingut's and Ollendorff's Spanish Grammar. New York: Ollendorff and
    Lockwood, 1865.
Last Poems by Elizabeth Barrett Browning. New York: James Miller, 1862.
Lessons in Greek Parsing, or Outlines of the Greek Grammar. Chauncey Goodrich. New Haven,
    Conn.: Durrie and Peck, 1831.
The Life and Voyages of Christopher Columbus. Washington Irving. New York: Harper, 1833.
Lucile. Owen Meredith. Boston: Ticknor and Fields, 1864.

New Spanish and English Dictionary. F. C. Meadows. Philadelphia: Thomas, Cowperthwait, 1832.

Ollendorff's New Method of Learning to Read, Write, and Speak the Spanish Language. M. Velasquez and T. Simonns. New York: D. Appleton, 1863.

Poems by Richard Henry Stoddard. Boston: Ticknor, Reid, and Fields, 1852.

Poetical Works of Lord Byron. Boston: Houghton Mifflin, 1863.

The Rambler. Vol. 2. Alexander Chalmers, ed. New York, 1811.

The Vision (Divine Comedy). Dante. Rev. Henry Francis Cary, trans. New York: D. Appleton, 1864.

The Works of Robert Burns. John Lockhart, ed. Hartford: Judd, Loomis, 1837.

Appendix 2. Injurious Insects in Illinois Today

As Illinois state entomologist, Stephen Forbes dealt with a dozen more important species of injurious insects during 1882–1917. Based on their abundance, damage caused, and economic effect, some eighty years later in Illinois these same species were rated in importance as follows:

Serious	codling moth, plum curculio
Moderate	corn bill-bug, Hessian fly, San José scale, white grubs
Moderate to low	chinch bug
Low	canker worm, corn root aphis, fruit bark beetle
Low to none	wheat stem maggot, wheat joint-worm

These ratings were made in 1997–98 by Dr. Rich Weinzierl, extension entomologist, Crop Sciences–Entomology, and Dr. Rob Wiedenmann, associate professional scientist, Entomology Division, Illinois Natural History Survey.

Appendix 3. Graduate Students and Research Associates

The passing on of Forbes's scientific legacy at Illinois was primarily through people who worked for him in the four university-state agencies that he directed over more then fifty years. Four of these individuals also took either doctoral (two) or master's (two) degrees with him.

Listed below are students who received graduate degrees and research associates who went on to do significant work elsewhere. Information is given about their specific connections with Forbes at Illinois, their most significant position(s) reached in their career, their areas of primary scientific competence, and the number of years they were associated with Forbes at Illinois. It should be noted that four of the individuals listed also did undergraduate work at Illinois during 1903–8, when Forbes still taught some undergraduate courses in entomology and zoology, including preparation and supervision of undergraduate theses. Nevertheless, the only undergraduate years that are counted in this listing are for Alfred Gross, who spent some undergraduate time as a research assistant to Forbes.

Davis, John J. (1885–1965): State Entomologist's Office; B.S. (Ill.) 1907; professor and head of entomology, Purdue University, 1920–56; scientific pest control; president of the Association of Economic Entomologists (1938); president of the Entomological Society of America (1932); 4 years.

Garman, Philip (1891–1968): B.S. (Ky.) 1913, M.S. (Ill.) 1914 with Forbes, Ph.D. (Ill.) 1916; senior entomologist, Conn. Agricultural Experiment Station, New Haven, 1954–60; Odonata, Acarina; 2 years.

Garman, W. Harrison (1858–1944): Illinois State Laboratory of Natural History; Johns Hopkins, 1881–82; D.Sc. (Ky.) 1912; professor of entomology, University of Kentucky, 1917–29; state entomologist of Kentucky, 1897–1929; economic entomology, ecology; president of the Association of Economic Entomologists (1905); 6 years.

Glasgow, Hugh (1884–1948): A.B. (Ill.) 1908, Ph.D. (Ill.) 1913 with Forbes; chief research entomologist, N.Y. Agricultural Experiment Station, Geneva, 1938–48; economic entomology; 5 years.

Glasgow, Robert D. (1879–1964): Illinois State Laboratory of Natural History, State Entomologist's Office, Illinois Natural History Survey; A.B. (Ill.) 1908, Ph.D. (Ill.) 1913 with Forbes; state entomologist of New York, 1928–49; physiology and ecology of insects; 9 years.

Gross, Alfred O. (1883–1970): Illinois State Laboratory of Natural History, Illinois Natural History Survey; A.B. (Ill.) 1908, Ph.D. (Harvard) 1912; Little Professor of Natural Sciences, Bowdoin College, 1950–53; comparative physiology of invertebrates, ornithology; 4 years.

Kofoid, Charles (1865–1947): Illinois Biological Station; A.B. (Oberlin) 1890, Ph.D. (Harvard) 1894; professor of zoology, University of California (Berkeley), 1910–36; marine plankton, protozoan morphology and taxonomy, tropical medicine (amoebiasis); National Academy of Sciences; 5 years.

Weed, Clarence M. (1864–1946): State Entomologist's Office; B.S. (Mich. St.) 1883, M.S. (Mich. St.) 1884, Sc.D. (Ohio St.) 1891; professor of zoology and entomology, New Hampshire State College, 1891–1904, instructor, to president, Massachusetts State Teachers College, Lowell, 1904–35; insect life histories, interrelations of flowers and insects, Phalangiidae; 2 years.

Woodworth, Charles W. (1865–1940): Illinois State Laboratory of Natural History; B.S. (Ill.) 1885, M.S. (Ill.) 1886 with Forbes; professor and head of entomology, University of California (Berkeley), 1913–40; economic entomology, pathology of insect poisons; 2 years.

Notes

ABBREVIATIONS AND FREQUENT SHORT TITLES IN NOTES

AV	Arthur Vestal Papers, University of Illinois Archives
Brown	D. A. Brown, *Grierson's Raid: A Cavalry Adventure of the Civil War* (Urbana: University of Illinois Press, 1954)
CO(I)	Chief's Office, Incoming correspondence, Illinois Natural History Survey, University of Illinois Archives
CO(O)	Chief's Office, Outgoing correspondence, Illinois Natural History Survey, University of Illinois Archives
CO(S)	Chief's Office, Subject file, Illinois Natural History Survey, University of Illinois Archives
CO(SOC)	Chief's Office, Natural History Society, Illinois Natural History Survey, University of Illinois Archives
DAB	*Dictionary of American Biography*
Forbes, *Memoriam*	E. B. Forbes, Stephen Alfred Forbes: His Ancestry, Education and Character, in *In Memoriam, Stephen Alfred Forbes, 1844–1930* (1930)
Forbes, "Raid"	S. A. Forbes, Grierson's cavalry raid. *Transactions of the Illinois State Historical Society* (oral paper, pt. 2, 1907)
FFL	E. F. Scott, Forbes Family Letters and Journals of Stephen Alfred Forbes, 1847–1870 (typescript, 1936)
Hays	R. G. Hays, *State Science in Illinois: The Scientific Surveys, 1850–1978* (Carbondale: Southern Illinois University Press, 1980)
Howard, *History*	L. O. Howard, *A History of Applied Entomology* (Smithsonian Miscellaneous Collection 84, 1930)
Howard, "Memoir"	L. O. Howard, Biographical Memoir of Stephen Alfred Forbes,

	1844–1930 (*Biographical Memoirs*, National Academy of Sciences 15, 1932)
INHS	Illinois Natural History Survey
Mallis	A. Mallis, *American Entomologists* (New Brunswick, N.J.: Rutgers University Press, 1971)
McIntosh	R. P. McIntosh, *The Background of Ecology: Concept and Theory* (Cambridge: Cambridge University Press, 1985)
Moores	R. G. Moores, *Fields of Rich Toil: The Development of the University of Illinois College of Agriculture* (Urbana: University of Illinois Press, 1970)
NAVR	United States National Archives, Veterans and Pension Records
NASAR	United States National Academy of Sciences, Archives
NHSSF	Stephen Forbes scientific papers, circulars, pamphlets, and reports, Illinois Natural History Survey Library
OR	The War of the Rebellion: A compilation of the official records of the Union and Confederate Armies (Washington, D.C.: Government Printing Office, 1880–1901) (series 1 only)
RBT	Reports of the Board of Trustees, Illinois Industrial University and University of Illinois
RC/	Author's interview
RCC/	Authors' correspondence
RSE	Reports of the State Entomologist of Illinois
RDLNH	Reports of the Director, State Laboratory of Natural History
SARNHS	State of Illinois Administrative Reports, Natural History Survey, University of Illinois Library
SF	Stephen Forbes
SF,IHS	Stephen Forbes Papers, Illinois Historical Survey
Solberg	W. V. Solberg, *The University of Illinois, 1867–1894* (Urbana: University of Illinois Press, 1968)
Starr	S. Z. Starr, *The Union Cavalry in the Civil War*, 3 vols. (Baton Rouge: Louisiana State University Press, 1979–85)
TSHOS	Transactions of the Illinois State Horticultural Society
VS	Victor Shelford Papers, University of Illinois Archives

1. BEGINNINGS, 1836–60

1. Ripley, "Settlement of Aux Plaines" (1936), 46–56; Robbins, *Landed Heritage* (1942), 63, 81.

2. Fulwider, *History of Stephenson County* (1910), 10, 57, 66, 153, 413; Pierce, *Forbes and Forbush Genealogy* (1892), 81; Ripley, "Settlement of Aux Plaines," 56, 60–61.

3. Pooley, *Settlement of Illinois* (1908), 554; land records, Stephenson County, Ill.; Morison, *Oxford History of the American People* (1972) 2:231, 354–55; Ripley, "Forbes Family" (n.d.), courtesy of R. M. Forbes; FFL xi; RCC/R. Forbes; land records,

Stephenson County Ill.; FFL viii; Forbes, *Memoriam,* 6; Scott, "Recollections" (1958), SF, IHS, III-B-3; Nettie Forbes to SF, FFL 463; Bethune, *Lays* (1847), 91.

4. FFL vii–ix, xii–xiii, xvi; Agnes Forbes to Henry Forbes, FFL 16, 18; land records, Stephenson County, Ill.; SF to Nettie Forbes, FFL 36; Morison, *Oxford History of the American People* (1972) 2:364–65; SF to C. Johnson 3/27/1917, SF, IHS, VII-A.

5. Beloit Academy Register, 1860–61; Agnes Forbes to SF, FFL 25, 30, 32; Henry Forbes to SF, FFL 27; Ramsey, *Jonathan Edwards* (1957), 33; SF, IHS, II-B1 and II-B2; Henry Forbes to SF, FFL 32; SF to Nettie Forbes, FFL 447; FFL viii.

2. THE WAR YEARS, 1861–65

1. Forbes, *The Illinois* (1911) 3(1):4; Forbes, "Old Soldiers" (1913), 4–5, SF, IHS, II-B8; SF to Nettie Forbes, FFL 44, 47, 56; SF to Frances Snow, FFL 51–53; SF to Nettie Forbes, FFL 59–60.

2. Morison, *Oxford History of the American People* (1972) 2:403, SF to Clara Forbes, FFL 36–37; SF, FFL 64–65; SF to Nettie Forbes, FFL 68–71; Horn, *Army of Tennessee* (1952), 144; Reece, *History of Seventh Cavalry* (1901), 102; Long, *Civil War* (1971), 185–96; Fiske, *Mississippi Valley* (1901), 101–2; SF to Flavilla Bliss, FFL 93–95; SF to Nettie Forbes, FFL 97–98.

3. Starr 3:59; Horn, *Army of Tennessee* (1952) 145; SF to Nettie Forbes, FFL 103, 105; Horn, *Army of Tennessee* (1952), 150; SF, FFL 116, 118–21.

4. SF, FFL 122–42, 146–49, 153–72; SF to Nettie Forbes, FFL 182–83, 191; Forbes, *The Illinois* (1911) 3(1):6.

5. SF to Nettie Forbes, FFL 194–95, 200–202, 225; Catton, *This Hallowed Ground* (1956), 211; Ballard, *Pemberton* (1991), 114, 137–40; Forbes, "Raid," 5–25; Brown, 211–12, 216–17, 222–23, 233–34; Starr 3:190–91, 195; Simon, *Papers of Ulysses S. Grant* (1979) 8:270; FFL 519.

6. Long, *Civil War* (1971), 391; SF to Nettie Forbes, FFL 222–25; OR 30(2) 741–43, 749; SF to Nettie Forbes, FFL 239–41; Henry Forbes to Flavilla Bliss, FFL 247; SF, IHS, VI-C (oversize); Mary Foster to SF, FFL 256; OR 31(1), 246–48; Henry, *Forrest* (1944), 13–22, 462–65.

7. Starr 3:372; SF to Nettie Forbes, FFL 270; FFL 519; OR 31(1) 621; Agnes Forbes to SF, FFL 268; Henry, *Forrest* (1944), 211–12; FFL 519.

8. Agnes Forbes to SF, FFL 257; Henry Forbes to Nettie Forbes, FFL 278–79; SF to Flavilla Bliss, FFL 284; Henry Forbes to SF, FFL 280–81; SF, FFL 312, 317–18, 321–24; Forbes, *The Illinois* (1911) 3(1):6, 9; FFL 520.

9. SF, FFL, 326–27, 332–33, 338–41, 348, 355–58, 365; SF to Nettie Forbes, FFL 340; OR 39(2) 142; Henry, *Forrest* (1944), 337–38, 342.

10. Henry, *Forrest* (1944), 352, 359; OR 45(1) 575–76; SF, FFL 382, 386–89, 392–94, 396–97; OR 45(1), 580; Hurst, *Nathan Bedford Forrest* (1993), 216–22; SF to Frances Snow, FFL 402; OR 45(1) 581; Henry, *Forrest* (1944), 381–84, 389; OR 45(1) 32, 340, 556, 575, 582–88; McMurray, *John Bell Hood* (1982), 169–70; FFL 519; Pierce, *Second Iowa Cavalry* (1865), 134–35; McPherson, *Battle Chronicle* (1989), 184; Sword, *Embrace*

an Angry Wind (1994), 122, 152–55, 241, 269–70; McDonough and Connelly. *Five Tragic Hours* (1983), 153–54.

11. McMurray, *John Bell Hood* (1982), 177; Starr 3:546; Horn, *Decisive Battle of Nashville* (1968), 21, 43–44, 59; OR 45(1) 37–39; OR 45(2) 189–90; OR 45(1) 551; Forbes, "The Old Soldiers," (1913), 7–8, SF, IHS, II-B-8; OR 45(1) 129, 435, 551, 563–64, 577, 589–91, 595, 722; Horn, *Decisive Battle of Nashville* (1968), 99–102.

12. OR 45(2) 194–95, 210; Horn, *Decisive Battle of Nashville* (1968), 118–19; OR 45(1) 552, 564, 577–78; Henry Forbes to Flavilla Bliss, FFL 430; SF to Nettie Forbes, FFL 420; OR 45(1) 134, 552, 578, 591, 688, 711; Horn, *Decisive Battle of Nashville* (1968), 144–45, 147; Horn, *Decisive Battle of Nashville* (1968), v–xiii; OR 45(1) 105; Horn, *Army of Tennessee* (1952), 417, 419; Horn, *Decisive Battle of Nashville* (1968), v; Wilson, *Under the Old Flag 2* (1912), 172; Caesar, *Battle for Gaul* 3:20–27, 70; OR 45(1), 567.

13. SF to Frances Snow, FFL 429; SF to Nettie Forbes, FFL 420–22; Longacre, *Union Stars to Top Hat* (1972), 197; Henry Forbes to Flavilla Bliss, FFL 434; Davenport, *Ninth Regiment Illinois* (1888), 174; SF to Nettie Forbes, FFL 461; Davenport, *Ninth Regiment Illinois* (1888), 177–79; SF, IHS, VI-B-6; Forbes, *The Illinois* (1911) 3(1):4; Agnes Forbes to SF, FFL 489, 505–6; Nettie Forbes to SF, FFL 490–91; SF to Nettie Forbes, FFL 492–94; Henry Forbes to Flavilla Bliss, FFL 498–99; SF to Henry Forbes, FFL 509, 511; Reece, *History of Seventh Cavalry* (1901), 104; Forbes, *The Illinois* (1911) 3(1):6–8, 10; Forbes, "The Old Soldiers"(1913), 5–6, SF, IHS, II-B-8.

3. MEDICINE AND TEACHING, 1866–70

1. Mumford, *Brown Decades* (1971), 2–5; Morison, *Oxford History of the American People* (1972) 2:499–501, 507; Henry Forbes to SF, FFL 512–13, 515–18, SF, IHS, III-B-2; Bonner, *Medicine in Chicago* (1991), 19–32, 34–35, 102, 229; Bynum, *Science and the Practice of Medicine* (1994), chap. 4, passim; Rothstein, *American Physicians* (1972), 196–97; Ludmerer, *Learning to Heal* (1985), 22–23; Rothstein, *American Medical Schools* (1987), 40–41; Cassedy, *Medicine in America* (1991), 75.

2. Bowman, *Good Medicine* (1987), 17; Bonner, *Medicine in Chicago* (1991), 47–48; Rothstein, *American Medical Schools* (1987), 55–58, 87, 92–93; Rush Medical College for 1867–68, 5–8, 10–16; SF, IHS, XII-8; Rush Medical College for 1866–67, 2–4.

3. SF to Frances Snow, FFL 521–22; Forbes, *The Illinois* (1911) 3(1):8–9; Rothstein, *American Medical Schools* (1987), 94, 99; Rush Medical College for 1866–67, 2–4; Norwood, "Medical Education" (1976), 487; SF to Nettie Forbes, FFL 522–23; Bowman, *Good Medicine* (1987), 1–4, 8, 17; Bonner, *Medicine in Chicago* (1991), 16–17, 179–81; Andreas, *History of Chicago* (1885) 2:523, 551–52; SF to Flavilla Bliss, FFL 524–25; Adams, *Doctors in Blue* (1996), 119–20; Cassedy, *Medicine in America* (1991), 41–42, 78.

4. Rush Medical College for 1867–68, 3–4; Bonner, *Medicine in Chicago* (1991), 58; SF to Nettie Forbes, FFL 526–28; SF to Flavilla Bliss, FFL 529; Cassedy, *Medicine in America* (1991), 59; Henry Forbes to SF, FFL 529–31; Horrell et al., *Land between the Rivers* (1973), 73, 78; SF to Flavilla Bliss, FFL 531–32; SF to Nettie Forbes, FFL 532–35.

5. SF to Flavilla Bliss, FFL 535–36, 538, Bonner, *Medicine in Chicago* (1991), 227; Federal Writers Project, *Illinois* (1947), 9; Horrell et al., *Land between the Rivers* (1973), 9, 11–13, 16, 18; Coleman, *Biology in the Nineteenth Century* (1971), 2–4; Cassedy, *Medicine in America* (1991), 43; Bynum, *Science and the Practice of Medicine* (1994), 6, 119; Hendrickson, "Forerunners" (1963), iii; DAB 10:229; G. Vasey to SF, SF, IHS, I-F-1.

6. Rush Medical College for 1867–68, 11, and 1866–67, 14; Rothstein, *American Physicians* (1972), 92–93; Ludmerer, *Learning to Heal* (1985), 16; Howard, "Memoir," xv, 3; SF to Flavilla Bliss, FFL 539–40; SF to Nettie Snyder, FFL 541–43, 547–52; SF, IHS, I-A-2-3.

7. SF to Nettie Snyder, FFL 554; SF, IHS, X; SF to Flavilla Bliss, FFL 555–59.

8. SF to Flavilla Bliss, FFL 562; SF to Nettie Snyder, FFL 561, 563–64; Mumford, *Brown Decades* (1971), 8; SF to Cornelia Winslow, FFL 565.

9. Forbes, "New Plants" (1870), 317–18, 352; Vasey, "Plants to name," (1870), 256; SF to Nettie Snyder, SF, IHS, I-A-2-1.

4. NATURAL HISTORY, 1871–84

1. SF to Flavilla Bliss, FFL 558; SF, IHS, ix–x; Agnes Forbes to Henry Forbes, FFL 7; Hendrickson, "Forerunners" (1963), 112, 115, 117–21; Cook and McHugh, *Illinois State Normal University* (1882), 26, 27, 236–37; Darrah, *Powell of the Colorado* (1969), 80, 146, 185; Marshall, *Grandest of Enterprises* (1956), 66, 78, 120–21, n.278; Forbes, "Former state natural history societies" (1907), 893–95; Solberg, 152; Clayton, *Illinois Fact Book* (1970), 40; CO (SOC), 1871.

2. Allen, *Naturalist in Britain* (1994), 162–63; Cook and McHugh, *Illinois State Normal University* (1882), 240–41; Nyhart, *Biology Takes Form* (1995), 90–94, 178–79; Forbes, *Chicago School* (1872) 5(54):313–16; Forbes, "Natural history" (1873) 367–69; G. Vasey to SF 4/1, 2, 22, 5/8, 10, 14, 20, 6/4, 6 and 14/1872; SF, IHS, I-F-1; SF to G. Vasey 5/14, 16, 6/4, and 6/1872; SF, IHS, I-F-1; DAB X (1964), 229; Darrah, *Powell of the Colorado* (1951), 181–82, 184–86, 204; Marshall, *Grandest of Enterprises* (1956), 128; Hays, 30; Catalogue, Normal University for 1872, 34; FFL 556; SF, IHS, III-B-2.

3. Clara Gaston to SF, 12/10/1872, SF, IHS, I-B-1-1; SF-IHS, III-B-2; SF to Nettie Snyder 2/26, 3/24, 5/4, 23, 6/28, and 8/11/1873, SF, IHS, I-A-2-2; Forbes, "Natural history" (1873), 363–73; Cook and McHugh, *Illinois State Normal University* (1882), 241, 243; Forbes, "Collecting and preserving specimens" (1874), 1–11, NHSSF 1974; Hendrickson, "Forerunners" (1963), 121; SF-IHS, III-B-1.

4. Ross, "Faunistic surveys" (1958), 129–31; Hendrickson, "Forerunners"(1963), 105–6 III; Mallis, 43, 50–52; Thomas, "Natural history survey" (1861); 663–65; Howard, *History*, 15; Decker, "Economic entomology" (1958), 106; Forbes, "Former state natural history societies" (1907), 895–96; Wright and Wright, "Agassiz's address" (1950), 503–6; Cook and McHugh, *Illinois State Normal University* (1882), 241–43.

5. Forbes, "Natural History School" (1876), CO (SOC); Forbes, "Natural history" (1874), 391–95; Dexter, "Historical sketch" (1973); SF to Flavilla Bliss 5/7 and 8/12/1875, SF, IHS, I-A-4-2; Forbes, "Former state natural history societies" (1907), 895, 897; Petulla, *American Environmental History* (1988), 205–8; Hays, 38;

Marshall, *Grandest of Enterprises* (1956), 153; T. Burrill to SF 1/4/1876, CO(I); Cook and McHugh, *Illinois State Normal University* (1882), 243.

6. Forbes,"Illinois Crustacea" (1876), 3–25; Hays, 42; Ross, "Faunistic surveys" (1958), 135–36; Bennett, "Aquatic biology" (1958), 163; D. Jordan to SF 12/17/1875, CO(I); Jordan, "Fishes of Illinois" (1878), 37–70; Forbes, "Illinois fishes" (1878), 71; McIntosh, 74, 76; S. Baird to SF 1/5 and 6/28/1877, CO(I).

7. *TSHOS* 9 (1876), 41; Forbes, *TSHOS* 10 (1877), 37–43; Wier, *TSHOS* 13 (1880), 60; T. G. Scott, "Wildlife research" (1958) 179, 184; McAtee, "Economic ornithology" (1933), 112; Forbes, *TSHOS* 12 (1879), 140–45; *TSHOS* 12 (1879), 146; Forbes, "State Laboratory," (1878), 5 NHSSF 1978.

8. Cook and McHugh, *Illinois State Normal University* (1882), 243–46; SF to J. Velie 1876, CO(O); Forbes, "State Laboratory," (1878), 6–11, 13, NHSSF; Rainger et al., *American Development of Biology* (1888), 58–63, 68, 70–71; Dunlap, *Nature and the English Diaspora* (1999), 139–44; Nyhart, *Biology Takes Form* (1995), 307–14, 320, 323–34; Allen, *Naturalist in Britain* (1994), 180–84, 190–92, 194–201, 206–11.

9. SF, IHS, I-B-1-1; Nettie Snyder to Flavilla Bliss, FFL 569–71; *Freeport Daily Bulletin*, 11/16/1877, 4; SF to Nettie Snyder 11/10/1877, 6/9 and 9/2/1878, SF, IHS, I-A-2-2; Forbes, "Food of Illinois fishes" (1878), 71–89; McIntosh, 32, 43, 48; Constitution and Record, CO(SOC), 1876–1881 folder; Hendrickson, "Forerunners" (1963), 123–24; Forbes, RDLNH 1880–82, 7, 10–12, NHSSF; Forbes, "Food of fishes" (1880), 2 (introd.); SF to Clara Forbes 4/22/1879, SF, IHS, I-B-1-1; SF to Clara Forbes, misc. letters, 1877–81; SF to Clara Forbes 8/24/1881, SF-IHS, I-B-1-2; SF to Clara Forbes, FFL 36–37; SF to Nettie Snyder 8/9/1879 and 7/10/1880, SF, IHS, I-A-2-2.

10. C. Thomas to SF 8/5/1880, CO(I); Forbes, "Food of birds" (1880), 121, 124–25; Forbes, "Food of fishes" (1880), 17, 81–86, 89–127, 133, 137–48, Forbes, "Food of the thrush family" (1881), 110; Forbes, "Food of the bluebird" (1880), 215, 233; T. Burrill to SF 12/7/1880, CO(I); McIntosh, 65, 74–76, 186.

11. Forbes, "Some interactions of organisms" (1880), 3–8, 11–17; McIntosh, "Pluralism in ecology" (1987), 328; McIntosh, 74.

12. McIntosh, 64; Forbes, "Food of Illinois fishes" (1878), 71–89; D. Jordan to SF 9/27/1880, CO(I); Forbes, "Food of young fishes" (1880), 66; Forbes, "Food of fishes" (1880) 20, 22–36, 41, 54, 61–64, and "Food of young fishes," 70–73, 75–79; Forbes, 2d Ann. Meeting, 2/8/1881, CO (SOC), 1876–81 folder.

13. Ward, "Stephen Alfred Forbes" (1930), 380; McAtee, "Economic ornithology" (1933), 111–12; Clara Forbes to Flavilla Bliss 9/13/1881, SF, IHS, I-C-4-2; Forbes RDLNH 1880–82, 1–2, 7–8, 11–12; T. Burrill to SF 2/26/1881, CO(I); S. Baird to SF 8/3/1881; CO(I); SF to Clara Forbes 8/24, 9/30, 10/17, and 30/1881, SF, IHS, I-B-1-2; SF to Nettie Snyder 8/15/1881, SF, IHS, I-A-2-2.

14. W. Barnard to SF 4/2/1882, CO(I); T. Burrill to SF 4/4 and 6/1882, CO(I); C. Butler to SF 4/6/1882, CO(I); Solberg, 108, 129, 242 (N66), 245–46; P. Earle to SF 4/30/1882, CO(I); C. Thomas to SF 5/1/1882, CO(I); SF, IHS IX-5; 11th RBT (1882) 223, 235–36; SF to Flavilla Bliss 6/30/1882, SF, IHS, I-A-4-2; D. Jor-

dan to SF 7/29/1882, CO(I); Jordan, *Days of a Man* (1922), 2:245; Forbes, RDLNH 1880–82, 3–4.

15. Forbes, "Regulative actions of birds" (1882), 3–5, 17, 20–21; J. Robinson to SF 5/18/1881, CO(I); Forbes, "Ornithological balance-wheel" (1882), 120–22, 128, and "Notes on economic ornithology" (1883), 58–60; SF to Nettie Snyder 3/31/1883, SF, IHS, I-A-2-3; Solberg, 108, 246, 253.

16. Forbes, "Food of smaller fresh-water fishes" (1883) 65, 70, 74–77, 90–94, and "First food of common white-fish," 97–108; Forbes, "Some entomostraca of Lake Michigan" (1882), 537–42, 640–49.

17. Solberg, 246, 248; T. Burrill to SF 3/10/1884, SF, IHS, V-A; 13th RBT (1887), 10, 19; 12th RBT (1885), 230, 234; P. Earle to SF 2/3/1884, CO(I) and 3/14/1884, SF, IHS, I-F-5; D. Jordan to SF 3/20/1884 and 4/21/1884, SF, IHS, I-F-2; E. Craig to M. Simmons 5/2/1952, NHSSF 1952; SF to K. Hewins 3/21/1925, NHSSF 1925; SF to Nettie Snyder 4/28/1884, SF, IHS, V-A; S. Peabody to SF 5/13/1884, SF, IHS, V-A; SF-IHS, II-B-15; E. Snyder to SF 6/18/1884, SF, IHS, V-A; S. Peabody to SF 6/19/1884, SF, IHS, V-A.

18. SF to Nettie Snyder 11/22/1884, SF, IHS, I-A-2-3; SF to S. Peabody 12/12/1884, in 13th RBT (1887), 10–11; Warrick, "Library"(1958), 212; E. James to SF 12/17/1884, SF, IHS, I-F-3; Hays, 39; *Illinois Laws* (1885), 23–24; Solberg 80, 226–27, 242, 245–49, 370; 13th RBT (1887), 18.

5. STATE ENTOMOLOGIST AND PROFESSOR, 1882–95

1. Solberg, passim; Catalog Ill. Industrial University 1884–85, 20, 70–77; Hays, 41–42; Forbes, "Symposium" (1909), 62, 65–66, NHSSF; Forbes, "Insect, farmer, teacher" (1915), 6–7, NHSSF; W. Nason to SF 5/18/1882, CO(I); 13th RBT (1887), 162–63; J. Comstock to SF 8/21/1884, CO(I); Decker, "Economic entomology" (1958), 104–7, 116; Davenport, "Natural scientists" (1958), 363–67; C. Thomas to SF 7/22/1882, CO(I); Garraty, *New Commonwealth* (1968), 51; Metcalf and Flint, *Destructive and Useful Insects* (1962), 480–81; Howard, "Memoir," 15, 29–53.

2. Forbes, "Kind of economic entomology" (1904), 3, 13, and passim, NHSSF; Forbes, "Insects affecting corn" (1883), 1–6, 12, 20–21, NHSSF; Forbes, "Contagious disease of caterpillars" (1884), 31; Forbes, "Ecological foundations" (1915), 2; Forbes, "Insect, farmer, teacher" (1915), 2, NHSSF; SF, IHS, II-A-10; Forbes, "Chinch-bug" (1886), 8, NHSSF; Forbes, "Hessian fly" (1890), 3, NHSSF; Forbes, "Chinch-bug in Illinois" (1887), 37–38, NHSSF.

3. Forbes, "Chinch-bug in Illinois" (1887), 27–28, 34–35, NHSSF; Forbes, "Kind of economic entomology" (1904), 1–16, NHSSF; Forbes, "Season's work" (1885), 117–18, 126–27; Forbes, "Ecological foundations" (1915), 11–12, Forbes, "Insect, farmer, teacher" (1915), 1–11; Ayars, "Publications and public relations" (1958), 204; Hays, 43.

4. Howard, *History* (1931), 426; Howard, USDA Yearbook (1888), 106; Forbes, "Arsenical insecticides" (1890), 310–12, 315–18, 322–25; Essig, *History of Entomology* (1931),

426; Forbes, "Codling moth" (1886) 111–12, 116–18; Metcalf and Flint, *Destructive and Useful Insects* (1962), 317; Forbes, "Arsenical poisons" (1887), 109–19.

5. Moores, 74–75, 98; Burrill and Forbes, "Agricultural experiment stations" (1885), 52–53, 60, 71; Solberg, 238, 241; Forbes, RDLNH (1888), 8–10, NHSSF and (1890), 2–3, 5, NHSSF; Howard, "Memoir," 36–41; Forbes, "Progress in economic entomology" (1891), 30, 32–34, NHSSF; Howard, *History*, 106–7.

6. NAVR; Howard, "Memoir," 20–21, 23–25; Alumni Morgue, Univ. Ill. Archives; Ward, "Stephen Alfred Forbes" (1930), 381; Mallis, 59; Scott, "Recollections" (1958), SF, IHS, III-B-3; SF to Flavilla Bliss 9/11/1886, SF, IHS, I-A-4-2; SF, IHS, III-B-1; Garraty, *New Commonwealth* (1968), 71–74, 310, 333–35; Clayton, *Illinois Fact Book* (1970), 392; Wiebe, *Search for Order* (1967), 18–43, chap. 3 passim; Forbes, "Insect, farmer, teacher" (1915), 5, 9, NHSSF; SF to Flavilla Bliss 9/11/1886, SF, IHS, I-A-4-2.

7. Solberg, 233–35, 249–50, 253, 303; Mills, "Stephen Alfred Forbes" (1964), 211; Howard, "Memoir," 20; Burrill, "Resignation" (1905), 2, SF, IHS, V-8, Pease, "Stephen Alfred Forbes" (1930), 545.

8. Forbes, "Some entomostraca" (1882), 824–25; Forbes, "Bacterial insect disease" (1891), 403; Decker, "Economic entomology" (1958), 116; 12th RSE (1883), 47–54; Forbes, "Insects affecting corn" (1883), 18–20, NHSSF; Forbes, "Contagious disease of caterpillars" (1884), 29–41; Forbes, "Contagious diseases of insects" (1886), 257–321; 13th RBT (1887), 294–301; Forbes, 19th RSE (1895), 22; Forbes, "Contagious insect diseases" (1888), 3–8, 11–12; T. Burrill to SF 11/27/1883, CO(I); Riley, "Contagious germs as insecticides" 17 (1883), 1169–70.

9. Forbes, 16th RSE (1890), 1–57, 18th (1894), 75–78, 19th (1895) 24–29 and 20th (1898), 106–9; Steinhaus, "Microbial control" (1956), 129, 144–45, 148; Snow, "Contagious diseases" (1895), 1–46; Cameron, "Insect pathology" (1873), 290–93; C. Riley to SF 1/15/1891, CO(I); Forbes, RDLNH (1892), 4, NHSSF; Forbes, "Field infection," *Spec. Bull. St. Entomol.* (1895), 1–4, NHSSF; Forbes, 13th RBT (1887), 300, and 17th (1894), 318–22; Decker, "Economic entomology" (1958), 116, 122; Forbes, "Insects affecting corn" (1883), 3–4, NHSSF; *Fosters' Daily Democrat*, Oct. 17, 1997, P. 12, Dover, NH; Steinhaus, *Insect Pathology* (1949), 525–36.

6. ECOLOGY AT THE CENTER, 1891–1917

1. Solberg, 147, 224, 263, 325, 328, 331–45, 349–51, 355–67; Moores, 98 (n.32); SF to Editor 11/18/1892, *Ottawa Naturalist* (1892) 7(8):133–35; Warrick, "Library" (1958), 212.

2. Howard, *History*, 109–11, 114; Forbes, "Presidential address" (1893), 61–70, NHSSF; McIntosh, 29–30, 48; Burdon-Sanderson, "Inaugural address" (1893), 465; Forbes, 19th RSE (1895), 16–17.

3. Howard, *History*, 55, 60, 119–24; Metcalf and Flint, *Destructive and Useful Insects* (1962), 705–7; Forbes, "San José scale" (1897), 413–15, 418–22, and "On the San José scale" (1915), 546–50; Forbes, "Work with the San José scale" (1899), 51–55, 57,

60–61, and "Season's campaign against the San José" (1898), 105–6, 109, 171–72; Forbes, "Nursery inspection" (1901), 174–75; Forbes, "Horticultural inspection act" (1901), 213–15, and "Nursery inspections" (1905), 110–11; Forbes, "Experiments with insecticides" (1902), 241–45, 255, 263–64; Forbes, 22nd RSE (1903), 96–97.

4. Moores, 116–18, 121, 136–38, 140–43, 240; Howard, "Memoir," 36–51; Forbes, "Symposium" (1909), 66–67; True, "History of agricultural experimentation" (1937), 168, 203, 207; Dupree, *Science in the Federal Government* (1957), 172; Howard, *History*, 102, 106–7, 540.

5. Howard, "Memoir," 29–51; Forbes, "Insects injurious to corn" (1905), 3–23, NHSSF; Forbes, "Indian corn plant" (1909), 286–87; Forbes, "Corn insects" (1897), 36–37; Forbes, "Kind of economic entomology" (1904), 4, 12, 15–16, NHSSF; Metcalf and Flint, *Destructive and Useful Insects* (1962), 501–3; Forbes, "Corn bill-bugs" (1902), 435–61, and "Insects injurious to Indian corn" (1905), 8 pp.

6. Catalogs, Ill. Industrial University and University of Illinois, 1884–1921, under Zoology and Entomology; Mallis, 446–47; SF 12/25/1923, SF, IHS, III-B-1; Henry Forbes to F. Snow 12/16/1863, FFL 266; Kaiser, "Wild eagle flight" (1975), 395, 406–7; SF to Clara Forbes 1/8/1905, SF, IHS, I-B-1-6; Forbes, *Memoriam*, 34, Burrill, "Resignation" 6/8/1905, SF, IHS, I- B; *Urbana Evening Post* 6/8/1905, p. 1; T. Clark, In Memoriam S. A. Forbes (1930), 39.

7. Forbes, "Nursery inspection" (1907), 126–27, 131–32; Forbes, "Entomologist's office" (1910), 140–41; Forbes, 25th RSE (1909), v–xi; Forbes, "Symposium" (1909), 66; Forbes, "Chinch-bug outbreak" (1916), 52–54, and "Chinch-bug situation" (1912), 7 pp.; Forbes, "Chinch-bug campaign" (1912), 17 pp., and "Chinch-bug situation" (1914), 3 pp.

8. Forbes, "Economic entomology" (1909), 1–2, 26–35; McIntosh, 67; RCC/Edward Smith, April 1994; Howard, "Memoir," 16–17.

9. Register, Univ. of Illinois under Entomology, 1909–21; Forbes, "Symposium" (1909), 55–56, 59, NHSSF; Howard, *History*, 76; Mallis, 220–21; Graduate degree data, courtesy of Richard Forbes from University of Illinois Archives and Rare Book Room; Forbes, Univ. of Illinois, "Entomological study" (1922), 4 pp., NHSSF; Solberg, 365–68.

10. SF, IHS, IX-13, oS; Clara Forbes to family 8/6/1912, SF, IHS, I-D; SF to Ernest Forbes 8/11/1912, SF, IHS, I-D; Forbes, *"Simulium"* (1913), 86–91; Roe, *History of Pellagra* (1973), 120-127.

11. Clara Forbes to family 8/12/1912, SF, IHS, I-D; Clara Forbes to Ernest Forbes 9/24/1912, SF, IHS, I-D.

12. True, "History of agricultural experimentation" (1937), 203, 228, 243–44; Howard, *History*, 165, 167; Howard, "Memoir," 49–51; Forbes, "Ecological foundations" (1915), 1–19; R. Wolcott to V. Shelford 3/27/1914, VS; Shelford, "Ecological Society of America" (1938), 165; Cowles, "Ecological Society," 9/20/1915, AV, Box 2; ESA, Announcement of foundation, 2 pp. AV, Box 2; F. Shreve to V. Shelford 3/4/1916, VS; Shreve, "The ESA" 4/1/1916; VS; Anon., *Journal of Ecology* 5 (1917), 119; Burgess, "Ecological Society of America" (1977), 4–5, 7.

7. AQUATIC SYSTEMS, 1880–93

1. Forbes, "Some interactions of organisms" (1880), 15, and "Food of fishes" (1980), 18–20; McIntosh, 74–76, 120, 195; taped interview of Victor Shelford with M. Brichford, 4/1/1965, VS, Box 2; Möbius, *Die Austern* (1877); Forbes, RDLNH (1883), 1–4; S. Baird to SF 8/10/1884, CO(I); Forbes, "Invertebrate animals" (1890), 479–87; Frey, *Limnology* (1963), 4, 16–17, 22–23, 36–37; Birge, "Plankton studies" (1898), 274–448; Birge, "Gases dissolved" 28 (1910), 1273–94; Forbes, "Invertebrate animals" (1890), 486; Juday, "Aquatic invertebrates" (1908), 10–16 and "Quantitative studies of bottom fauna" (1921), 461–93.

2. Forbes, RDLNH (1888) 2, 8, NHSSF; Forbes, "interactions of organisms" (1880), 3–17, 18–19; Howard, "Memoir," 2; F. Brendel to SF 6/18/1886, CO(I); Forbes, "Lake as a microcosm" (1925), 537–50; McIntosh, 59, 95, 120–21, 123–24, 253, 255; Hutchinson, "Prospect before us" (1963), 689; Lewontin, "Corpse in the elevator," Jan 20 (1983), 34–37. The Peoria Scientific Association (1875–1900) was founded by three physicians with ten other charter members, including Emma Smith, assistant state entomologist. Original subjects of interest were scientific but by 1886 included history, literature, and education. By 1887 membership had increased to more than a hundred, a total of 167 papers had been read, and the only bulletin ever published by the association appeared, which included Forbes's "Microcosm" paper (Pollak, "Peoria Scientific Association" [1963], 9 pp.).

3. Forbes, RDLNH (1888), 2–3, NHSSF; Garman, "Animals of the Mississippi" (1890), 123–84; Carlander et al., "Mid-continent states" (1963), 324; Bennett, "Aquatic biology" (1958), 163–64; Forbes, "Lake as a microcosm" (1925), 538–39; Schneider, "Enclosing the floodplain" (1996), 70–72, 80–81, 84–85; Forbes, "Food relations of freshwater fishes" (1888), 476–77, 479–84, 488, 490, 502–4; Forbes, "Mississippi" (1888), 1–3, 16–17.

4. Forbes, RDLNH (1890), 1, 5, NHSSF; Forbes, "Lake Superior Entomostraca" (1891), 701–18 passim; Forbes, "Some entomostraca" (1882), 540–41; Birge, "Crustacea from Madison" (1892), 379–98; Forbes, "Aquatic invertebrate fauna" (1893), 207; Pennak, "Rocky Mountain states" (1963), 365; R. Rathbun to SF 9/27/1890 and 10/12/1891, CO(I); SF to R. Rathbun 7/9/1890, CO(I); D. Jordan to SF 6/9/1890, CO(I); SF to Elwood Hofer 6/12/1890, CO(O); SF to Gov. J. Fifer 6/23/1890, CO(O); Forbes, "Aquatic invertebrate fauna"(1893), 207–8, 211, 214, 224.

5. SF to Nettie Snyder 7/20/1890, SF, IHS, I-A-2-3; SF to Clara Forbes 7/21/1890, SF, IHS, I-B-1-4; Forbes, "Aquatic invertebrate fauna" (1893), 208–10, 215–28, 234–35; SF, IHS, Misc. family letters, July 1890.

6. Forbes, "Aquatic invertebrate fauna" (1893), 210, 220, and passim; Anderson, "Trout and invertebrate species" (1980), 635–41; SF to Bertha Forbes 8/30/1890, SF, IHS, I-B-1-4; SF to R. Rathbun 7/16 and 8/1/1891, CO(O); R. Rathbun to SF 7/18 and 7/27/1891, CO(I); Forbes, "Aquatic invertebrate fauna" (1893), 212–13, 236–39; Pennak, *Fresh-water Invertebrates* (1989), 389, 392, 432, 436; SF to Clara Forbes 9/6/1891, SF, IHS, I-B-1-4.

In 1980, for example, R. S. Anderson of the University of Calgary, Alberta, Canada, provided comparisons of zooplankton community structure and predators for 320 fishless and fish-bearing small mountain lakes near the continental divide in the Canadian Rockies. The species complements sketched for Forbes's Yellowstone lakes for the most part fit into the wide array of those Anderson observed in these Canadian lakes. Changes in community structure after introduction of fish into mountain lakes would involve the roles of predation and competition affecting both fish and invertebrates. For example, a common result of adding rainbow or brook trout is fewer diaptomid copepods, whose eggs are eaten by more abundant cyclopoid copepods.

7. 16th RBT (1892), 244; Forbes, RDLNH (1892), 2–3, NHSSF; R. Rathbun to SF 4/18, 5/2, 5/17, 6/2 and 6/11/1892, CO(I); M. McDonald to SF 6/11 and 10/27/1892, CO(I); SF to M. McDonald 6/24/1892, CO(O); SF to R. Rathbun 7/14/1892, CO(O); A. Verrill to SF 9/21/1892, Co(I); SF, IHS, IX-8; T. Bean to SF 12/31/1892, CO(I); SF to H. Armsby 1/12/1893, CO(O); SF, IHS, II-A-6; SF to Clara Forbes 1/8/1893, SF, IHS, I-B-1-5.

8. Forbes, "Aquarium" (1894), 143–58; Badger, *Great American Fair* (1979), xi–xii, 109, 125, 131; Anon., *Gems* (1894), n.p.; SF to Clara Forbes 2/12/1893, SF, IHS, I-B-1-5; Clara Forbes to Mother Gaston N.D., SF, IHS, I-C-4-3; SF to Clara Forbes 4/5/1893 (2 letters), SF, IHS, I-B-1-5; SF to Clara Forbes 4/6, 7 and 12/1893, SF, IHS, I-B-1-5; Forbes, "Aquarium" (1894), 143; M. McDonald to SF 6/5/1893, CO(I); Forbes, RDLNH, in 17th RBT (1894), 308–10; M. McDonald to SF 10/27/1892, CO(I).

8. THE ILLINOIS BIOLOGICAL STATION, 1894–1915

1. Forbes, "Biological investigations" (1910), 1, 5; Solberg, 366–68; Forbes, RDLNH, in 17th RBT (1894), 311–14; 18th RBT (1896), 114–15; Wilder, "University of Illinois Biological Station" (1896), 332; Pillsbury, 20th RBT (1901), 75–76; Forbes, "Biological Experiment Station" (1896), 323, 325, NHSSF; SF to J. Reighard 3/15/1894, CO(O); D. Jordan to SF 9/26/1894, CO(I); M. McDonald to SF 10/9/1894, CO(I); Anon., *Natural Science* (April 1896), n.p.; SF to A. Draper 11/21/1894, CO(O); SF to J. Armstrong 4/9/1896, CO(O); misc. letters, 1894, SF to Birge, Goode, and Packard, CO(O); Forbes, RDLNH, in 17th RBT (1894), 312.

2. Forbes, "Effects of stream pollution" (1928), 279; Forbes, RDLNH, in 17th RBT (1894), 314–17; Forbes, "Biological Experiment Station" (1896), 303, 305, 307–8, 310–13, NHSSF; Havera and Roat, "Forbes Biological Station" (1989), 6–8; *Natural Science* (April 1896).

3. R. Rathbun to SF 5/3 and 9/1894, CO(I); M. McDonald to SF 5/11/1894, CO(I); SF to M. McDonald 5/24/1894 and A. Talbot to M. McDonald 7/24/1894, in U.S. House Rep. Doc. No. 96, 53rd Congress; Bocking, "Forbes, Reighard, and aquatic ecology"(1990), 473–74, 482–83; E. Birge to SF 5/22/1895, CO(I); Forbes, "Biological Experiment Station" (1896), 309, 313, 315; NHSSF: Ko-

foid, "Fresh-water biological stations" (1898), 391; Wilder, *The Illini* 25, no. 20, Feb. 28, 1896, 318–19; Forbes, RDLNH (1898), 5, 16–19, NHSSF; Kofoid, "Plankton studies" (1897), 11–12, 213.

4. Kofoid, "Plankton studies" (1897), 12–13, 16–17, 19–20; Kofoid, "Plankton method" (1897), 832; Mills, *Biological Oceanography* (1989), 35–36, 133–37; Lohmann, "Ueber das Fischen mit Netzen" (1908), 127–370, and 10:129–370; Hays, 16–17, 20–21, 52–55; Forbes, RDLNH (1896), 299–301, NHSSF; Forbes, "Biological Experiment Station" (1896), 315, 319, 321–23, NHSSF; Forbes, RDLNH (1898), 23–25, 29–31, NHSSF.

5. D. Jordan to SF 9/14/1895, CO(I); R. Rathburn to SF 4/3/1896, CO(I); Bocking, "Forbes, Reighard, and aquatic ecology" (1990), 484–85; First District Congressman from Illinois to Pres. Wm. McKinley 5/1/1897, SF, IHS, I-F-5; G. Bowers to SF 4/23/1898, CO(I); SF to Clara Forbes 7/28/1897, SF, IHS, I-B-1-5.

6. Forbes, RDLNH (1901), 3–7, NHSSF; Forbes, "Biological investigations" (1910), 3, NHSSF; Pillsbury, 19th RBT (1898), 120.

7. Forbes, RDLNH (1898), 5, NHSSF; E. Birge to SF 1/21/1907, CO(I); Gunning, "Illinois" (1963), 170; Kofoid, "Plankton studies" (1908), 3–4, 10, 12–16, 312–13; Kofoid, "Plankton method" (1897), 829–32; Kofoid, "Fresh-water biological stations"(1898), 402; Kofoid, "Illinois River plankton" (1905), 233–34; Goldschmidt, *Biographical Memoirs of Kofoid* 26 (1951), 121–51; NASAR, 1922 class.

8. Forbes, RDLNH (1901), 7–8; Forbes, "Biological investigations" (1910), 3–4, NHSSF; Howard, "Memoir," 11–12, 14, 20–23; Scott, "Recollections" (1958), SF, IHS, III-B-3; Ayars, "Publications and public relations" (1958), 208; S. Batzli, *Preserv. & Conserv. Assoc. Newsletter* 15(3), 2 pp.

9. Forbes, "Biological investigations" (1910), 2–6, 9, 14; Forbes, "Biological Experiment Station" (1896), 307, 314–21, 325; Forbes, RDLNH (1901), 8, and passim, NHSSF; Forbes, "Illinois Biological Station" (1911), 226–27; Hays, 48, 56; Starrett, "Man and the Illinois River" (1972), 145–47; Forbes and Richardson, "Ilinois River biology" (1919), 139; Bellrose et al., "Waterfowl populations"(1979), 3–4; Bellrose et al., "Fate of lakes" (1983), 11.

10. *American Men of Science*, 4th ed. (1927), 817; Ross, "Faunistic surveys" (1958), 131; R. Richardson to SF 7/29/1903, CO(I); SF to R. Richardson 7/29 and 8/14/1903, CO(O); Forbes, RDLNH (1901), 3–6, NHSSF; Forbes, "Distribution of Illinois fishes" (1907), 273–83, 285–87, 295–97; McIntosh, 65; Forbes, "Fresh water fishes" (1914), 16–17, NHSSF; B. W. Evermann to SF 3/22/1909, CO(I); Burr, "Fishes of Illinois" (1991), 418; Smith, *Fishes of Illinois* (1979); Forbes and Richardson, *Fishes of Illinois* (1908), xii, xiv, lxxvii, lxxviii.

11. Forbes, "Streams pollution" (1921), 1–2; Bellrose et al., "Fate of lakes" (1983), 24; Forbes and Richardson, "Illinois River biology" (1919), 146–50, 153–54; Forbes, "Biological and chemical conditions" (1913), 166–67; Richardson, "Small bottom and shore fauna" (1921), 409, 433, 462–64 and passim; Forbes, "Fresh water fishes" (1914), 5–6; NHSSF; *American Men of Science*, 4th ed. (1927), 817; Schneider, "Enclosing the floodplain" (1996), 77–84.

12. Forbes, "Chemical and biological investigations" (1912), 5–6; Forbes, "Biological and chemical conditions" (1913), 168–70; Bennett, "Aquatic biology" (1958), 168; Kolkwitz and Marsson, "Oekologie" (1909), 126–52; Wilhm, "Biological indicators of pollution" (1975), 376; Forbes and Richardson, "Illinois River biology" (1919), 148; SF to T. Condit 11/5/1912, CO(S), Box 1, Illinois River.

13. NAVR; SF to Clara Forbes 7/22/1908, 11/14 and 26/1908, 12/4/1910, SF, IHS, I-B-1-6, SF to R. Richardson 7/11 and 9/10/1912, CO(S), Box 1; CO(S), Box 1, 1912, U.S. Bureau of Fisheries; SF to Clara Forbes 12/13/1910, SF, IHS, I-B-1-6; RCC/T. Scott 10/17/1986; RCC/Richard Forbes 3/10/1998.

14. Schneider, "Enclosing the floodplain" (1996), 80, 84, 86; Mills. et al., "Man's effect" (1966), 8; CO(S), Box 2, 1913–14, Illinois River Conference; Richardson, "Bottom and shore fauna" (1921), 46; Richardson, "Small bottom and shore fauna" (1928), 402; Richardson, "Bottom fauna" (1921), 363–66, 373–76; Forbes, "Biological investigations" (1910), 14; Schneider (2000), 696–98.

9. CHIEF OF THE SURVEY, 1917–23

1. Hays, 87–89, 93, 95, 97, 103–4; SARNHS, 1 (1918), 606; Ross, "Faunistic surveys" (1958), 132; SF to F. Shepardson 12/4/1917 and 4/26/1918, CO(S), Box 10, State DRE, Univ. of Illinois and INHS problems; Report to Board of Nat. Resources and Conserv., 10/16/1920, 2, INHS; SF to R. Richardson 10/18/1915, CO(S), Box 6; Alvord and Burdick, Report Rivers and Lakes Comm., 1915; Forbes and Richardson, "Illinois River biology" (1919), 156.

2. SF to R. Richardson 5/24/1917, CO(S), Box 7; SF to Clara Forbes 7/9/1917, 11/22/1918, SF, IHS, I-B-1-8; E. Forbes to L. Howard 10/16/1930, SF, IHS, VII-B-1; NASAR.

3. Richardson, "Bottom and shore fauna" (1921), 33–75; *Peoria Journal* 10/11/1919, n.p., CO(S), Box 13, Richardson; Hays, 96; E. Bartow to SF 11/7/1919, CO(S), Box 13, State DRE, Water Survey; SF to F. Shepardson 2/21 and 3/1/1920, and SF to W. Sackett 4/15/1920, CO(S), Box 13, State DRE, Rivers; SARNHS 3 (1920), 784–89, 5 (1922), 913; SF to R. Barney, Summer 1921, CO(S), Box 16, U.S. Fairport Biol. Sta.; SF to Clara Forbes 8/1/1920, SF, IHS, I-B-1-8.

4. Forbes, "Humanizing of ecology" (1922), 89–92; McIntosh, 296–97, 301–5.

5. Richardson, "Illinois land birds" (1921), 33, 46, 75; Forbes, "Sewage pollution" (1924), 35; Richardson, "Small bottom fauna" (1925), 327, 329, 332, 336–37, 345–46; Richardson, "Bottom fauna" (1928), 398–99, 402; Bellrose et al., "Waterfowl populations" (1979), 4; Richardson, "Illinois River bottom fauna" (1925), 391–422; Mills et al., "Man's effect" (1966), 15; Forbes and Richardson, "Illinois River biology" (1919), 154; Bennett, "Aquatic biology" (1958), 169; SF to L. Pease 4/17/1924, CO(S), Box 18, Sanitary District; SF to H. O'Malley 4/17/1924, CO(S), Box 18, U.S. Bureau of Fisheries.

6. Richardson, "Illinois River bottom fauna" (1925), 413; Richardson, "Bottom fauna" (1928), 406–13, 469; Richardson, "Small bottom fauna" (1925), 333; Havera and Roat, "Forbes Biological Station" (1989), 6; Hays, 100, 108; SF to R. Barney

10/12/1923, CO(S), Box 18, U.S. Biological Station; Forbes, "Biological survey" (1928), 278–84; SARNHS 7 (1924), 1253, 9 (1926), 970–71, 10 (1927), 1165–66, 11 (1928), 1289; Bennett, "Aquatic biology" (1958), 169; Forbes, "Effects of stream pollution" (1928), 12, NHSSF.

7. Forbes and Gross, "Orchard birds" (1921), 1–2; Forbes and Gross, "Illinois land birds" (1922), 188, 193–201, 205–6, 209–10, 214; Forbes and Gross, "Illinois land birds" (1923), 397–400, 404–12, 415, 429–36; McIntosh, 160.

8. Forbes, "Recent forest survey" (1919), 103–10; F. Shepardson to SF 12/23/1918, CO(S), Box 12, State DRE, Forestry and Rivers; Forbes, SARNHS 2 (1919), 519; Forbes and Miller, "Concerning a forestry survey" (1920), 3, 6; Federal Writers Project, *Illinois* (1947), 8; SF to F. Shepardson 3/4/1919, CO(S), Box 12, State DRE, Forestry & Rivers; SARNHS 4 (1921), 804–7; Chapman and Miller, "Forestry survey of Illinois" (1924), vii.

9. Forbes, *Illinois Blue Book for 1925–26* (1926), 470, 472, and *for 1929–30* (1930), 398; Chapman and Miller, "Forestry survey of Illinois" (1924), 46–172; Forbes, SARNHS, 9 (1926), 967–69, 10 (1927), 1163–64, 11 (1928), 1289, 12 (1929), 1044–45; SF to J. Peters 1/18/1926, CO(S), Box 20, U.S. Forest Service; Van Cleave, "Forbes as scientist" (1930), 282; Hays, 50–51, 191–94; Forbes, RDLNH, In 17th RBT (1894), 311; Bocking, "Forbes, Reighard, and aquatic ecology" (1990), 497 and passim.

10. THE PATH BEFORE HIM, 1923–30

1. SF, IHS, III-B-2; Forbes, *Memoriam*, 10–11; Farquharson, *Marcus Aurelius: Meditations* (1944), 19 (IV. 4).

2. RCC/T. Scott 10/17/1986; Scott, "Recollections" (1958), SF, IHS, III-B-3; SF to J. Scott 6/29/1924, SF, IHS, I-F-5; SF to Clara Forbes 8/16/1927 SF, IHS, I-B-1-10; Pease, "Stephen Alfred Forbes" (1930), 547; Forbes, *Memoriam*, 14; SF to Clara Forbes 8/8/1926 and 7/23/1927, SF, IHS, I-B-1-10.

3. SF to F. Scott 2/5/1927 and to Ethel Scott 1/29 and 6/3/1927, SF, IHS, I-B-3-4; RCC/T. Scott 10/17/1986; SF to Clara Forbes 8/12/1927, SF, IHS, I-B-1-10; D. Jordan to SF 8/6/1927, SF, IHS, I-F-2; SF to Clara Forbes 8/15/1927, SF, IHS, I-B-1-10.

4. Hays, 105–8; Forbes, "Natural History Survey" (1930), 11 pp., NHSSF; Anon., *Ill. Alum. News* 8(7) (1930), 279; "In Appreciation to Stephen A. Forbes" 3/29/1928, courtesy of Richard Forbes; Alumni Morgue, Univ. Ill. Archives.

5. R. Kelso to Ernest Forbes 4/5/1936, courtesy of Richard Forbes; SF to Ethel Scott 7/4, 8/3 and 31, 10/12 and 19, 11/24 and 12/25/1929, SF, IHS, I-B-3-5; SF to Winifred Forbes 12/30/1929, SF, IHS, I-B-4-2; SF, IHS, VI-A; SF to Ethel Scott 1/2/1930, SF, IHS, I-B-3-5; Alumni Morgue Univ. Ill. Archives, Urbana, Ill., newspaper 1/26/1930 np; SF to Winifred Forbes 1/30 and 2/2/1930, SF, IHS, I-B-4-2.

6. SF to Ethel Scott 2/10, 2/15, and 3/1/1930, SF, IHS, I-B-3-5; SF to Winifred Forbes 2/19/1930, SF, IHS, I-B-3-5; Pease, "Stephen Alfred Forbes" (1930), 548; Alumni Morgue, Univ. Ill. Archives; Forbes, *Memoriam*, 15; E. Johnson, *In Memoriam*, 20–21.

7. SF, IHS, VI-B; Metcalf, "Obituary" (May 1930), 178.

8. Anon. *Ill. Alum. News* 8(7) (April 1930), 278–82; Forbes, *In Memoriam* (1930), passim. Stephen was buried beside Clara in Roselawn Cemetery on the Illinois campus. He and Clara left an estate of $81,000, mostly bonds and stocks in the form of a trust for their four children (SF, IHS, VI-B, and C-21).

Bibliography

All the book, paper, and article titles listed below were employed in writing this book. Nearly two-thirds are primary sources. I also selected 445 letters from Forbes's correspondence with his family members, friends, scientific colleagues, and officials. With the addition of the various records, references, and personal papers also listed, the reader holds a book heavily based on primary sources. The remaining secondary sources were helpful for supplying background.

I based the discussion of Forbes's scientific work on 122 (about 30 percent) of his papers published between 1870 and 1930, including 70 percent (39 of 56) from among those papers considered by Forbes as his more important scientific publications from 1876 to 1916. I used this block of material to show how Forbes went about his work, how he developed and tested his ideas, and why these papers were noteworthy.

There are only three sizable writings on Stephen Forbes: first, Leland O. Howard's (1932) splendid fifty-four-page biographical memoir, including a chronology and a Forbes bibliography; second, Stephen Bocking's (1990) fine comparison of Forbes's and Jacob Reighard's early work in aquatic ecology in the Great Lakes region; and third, a Wisconsin doctoral dissertation by Robert Lovely (1995).

The unpublished manuscripts listed reside in the Stephen Forbes Collection, Illinois Historical Survey, unless otherwise indicated.

UNIVERSITY RECORDS AND PERSONAL PAPERS

1. University of Illinois Archives
   Alumni morgue
   Board of Trustees, University of Illinois, reports, 1887–1901
   Catalogue and register of the Illinois Industrial University and the University of Illinois, 1884–1921

Natural History Survey
—Chiefs' Correspondence: Incoming, 1871–1909; Outgoing, 1877–1911
—Economic Entomology, 1899–1909
—Illinois Natural History Society, 1858–60, 1871, 1877–85
—Subject file (correspondence), mostly 1912–31, State Laboratory of
    Natural History, State Entomologist's Office, and Illinois Natural
    History Survey
—Victor E. Shelford Papers
—Arthur J. Vestal Papers
2. University of Illinois–Illinois Historical Survey
    Stephen A. Forbes Collection, 900 letters plus 320 other items, 1842–1961, but
    mostly 1861–1930, and Forbes family letters and war journals of Stephen Forbes,
    1847–70
3. Illinois Natural History Survey, Library
    Stephen Forbes scientific papers, reports, circulars, and pamphlets, 1870–1930
4. University of Illinois Library
    State of Illinois administrative reports, Dept. of Registration and Education,
    Natural History Survey

OTHER RECORDS AND REFERENCES

1. Annual Announcement of Rush Medical College, Chicago, 1866–67, 1867–68
2. Beloit College and Academy Register, Beloit, Wis., 1860–61
3. Bowdoin College Special Collections, Brunswick, Maine
4. Bradley University Special Collections, Peoria, Ill.
5. Catalogue of the Illinois State Normal University, 1872
6. *Dictionary of American Biography*
7. Land records, Stephenson Co., Ill., 1842–54, 1861
8. A compilation of the official records of the Union and Confederate Armies,
    series I, Washington, D.C.
9. U.S. National Archives—Military and Pension Records
10. National Academy of Sciences—Archives

MANUSCRIPTS AND A DISSERTATION

Burrill, T. 1905. "The resignation of Dean S. A. Forbes."
Forbes, S. A. 1913. "The old soldiers of the Civil War."
Forbes, S. A. 1916. "List of more important scientific contributions."
Hottes, C. 1931. "Early contributions to economic entomology."
Lovely, R. A. 1995. Mastering Nature's Harmony: Stephen Forbes and the Roots of
    American Ecology. Ph.D. dissertation, University of Wisconsin, Madison. (In Illi-
    nois Natural History Survey Library.)
Ripley, V. S. n.d. "The Forbes family." (With Richard E. Forbes.)
Scott, E. F. 1958. "Recollections" of Stephen Forbes.

Thompson, D. H. 1935. "Notes on Robert E. Richardson." (With Thomas E. Rice, technical assistant, Illinois Natural History Survey.)

BOOKS, PAPERS, AND ARTICLES

Adams, G. W. 1996. *Doctors in Blue*. Baton Rouge: Louisiana State University Press.

Allen, D. E. 1994. *The Naturalist in Britain: A Social History*. Princeton, N.J.: Princeton University Press.

Alvord, C. W. 1920. *The Illinois Country 1673–1818*. Springfield: Illinois Centennial Commission.

Alvord, J. W., and C. B. Burdick. 1915. Report of the Rivers and Lakes Commission on the Illinois River and its bottomlands, with reference to the agriculture and fisheries and the control of floods. Springfield.

Anderson, R. S. 1980. 56. Relationship between trout and invertebrate species as predators and the structure of the crustacean and rotiferan plankton in mountain lakes. In *Evolution and Ecology of Zooplankton Communities*, edited by W. C. Kerfoot, 635–41. Hanover, N.H.: University Press of New England.

Andreas, A. T. 1884. *History of Chicago: From the Earliest Period to the Present Time*. Vol. 1, ending with the year 1857. Chicago: A. T. Andreas. New York: reprint, Arno Press, 1975.

————. 1885. *History of Chicago: From the Earliest Period to the Present Time*. Vol. 2, 1857–71. Chicago: A. T. Andreas. New York: reprint, Arno Press, 1975.

Anon. 1837. Illinois in 1837: A sketch and agricultural production. Suggestions to emigrants, and a letter on the cultivation of the prairies. Philadelphia: S. A. Mitchell Publ. (Amer. Culture Series, vol. 3, reel 616.12)

————. 1894. *Gems of the World's Fair and Midway Plaisance*. Philadelphia: Historical Publishing.

Applebaum, S. 1980. *The Chicago World's Fair of 1893: A Photographic Record*. New York: Dover.

Ayars, J. S. Publications and public relations. In Mills et al. 1959. "A Century of biological research." *Ill. Nat. Hist. Surv. Bull.* 27, art. 2, 202–9.

Badger, R. 1979. *The Great American Fair: The World's Columbian Exposition and American Culture*. Chicago: Nelson Hall.

Ballard, M. B. 1991. *Pemberton: A Biography*. Jackson: University Press of Mississippi.

Bellrose, F. C., S. P. Havera, F. L. Paveglio Jr., and D. W. Steffeck. 1983. The fate of lakes in the Illinois River valley. *Ill. Nat. Hist. Surv. Biol. Notes* no. 119. 27 pp.

Bellrose, F. C., F. V. Paveglio Jr., and D. W. Steffeck. 1979. Waterfowl populations and the changing environment of the Illinois River valley. *Ill. Nat. Hist. Surv. Bull.* 32, art. 1, 1–54.

Bennett, G. W. 1958. Aquatic biology. In Mills et al. 1959. "A Century of biological research." *Ill. Nat. Hist. Surv. Bull.* 27, art. 2, 163–78.

Benson, K. R. 1988.. From museum research to laboratory research: The transformation of natural history into academic biology. Chap. 2 in *The American Development of Biology*, edited by R. Rainger, K. R. Brown, and J. Maienschein, 49–83. Philadelphia: University of Pennsylvania Press.

Bethune, G. W. 1847. *Lays of Love and Faith*. Philadelphia: Lindsay.

Birge, E. A. 1892. Notes and list of Crustacea from Madison, Wisconsin. *Trans. Wis. Acad. Sci. Arts Lett.* 8:379–98.

————. 1898. Plankton studies on Lake Mendota. II. The crustacea of the plankton from July 1894 to December 1896. *Trans. Wis. Acad. Sci. Arts Lett.* 11:274–448.

————. 1910. Gases dissolved in the waters of Wisconsin lakes. *Bull. U.S. Bur. Fish.* 28:1273–94.

Birge, E. A., H. F. Nachtrieb, and F. Smith. 1901. State natural history surveys. *Science* 13 (328):563–68.

Black, A. 1870. Ernest Browning Forbes—A biographical sketch. *J. Nutr.* 1015–22.

Bocking, S. 1990. Stephen Forbes, Jacob Reighard, and the emergence of aquatic ecology in the Great Lakes region. *J. Hist. Biol.* 23(3): 461–98.

Bode, C., ed. 1967. *American Life in the 1840s*. New York: New York University Press.

————. 1972. *Midcentury America: Life in the 1850s*. Carbondale: Southern Illinois University Press.

Bogue, A. G. 1963. *From Prairie to Corn Belt: Farming on the Illinois and Iowa Prairies in the Nineteenth Century*. Chicago: University of Chicago Press.

Bonner, T. N. 1991. *Medicine in Chicago, 1850–1950*. Urbana and Chicago: University of Illinois Press.

Boritt, G. S., ed. 1992. *Why the Confederacy Lost*. New York: Oxford University Press.

Bousfield, E. L. 1958. Fresh-water amphipod crustaceans of glaciated North America. *Can. Field. Nat.* 72:55–113.

Bowman, J. 1987. *Good Medicine: The First 150 Years of Rush-Presbyterian–St. Lukes Medical Center*. Chicago: Chicago Review Press.

Branch, E. D. 1934. *The Sentimental Years, 1836–1860*. New York: Appleton-Century.

Brown, D. A. 1954. *Grierson's Raid: A Cavalry Adventure of the Civil War*. Urbana: University of Illinois Press.

Burdon-Sanderson, J. S. 1893. Inaugural address. *Nature* 48:464–72.

Burgess, R. L. 1977. The Ecological Society of America: Historical data and some preliminary analysis. In *History of American Ecology*, edited by F. N. Egerton, 1–24. New York: Arno Press.

Burland, R., and E. Burland. 1974. *A True Picture of Emigration*. Secaucus, N.J.: Citadel Press.

Burr, B. M. 1991. The fishes of Illinois: An overview of a dynamic fauna. In *Our living heritage: The biological resources of Illinois*, edited by L. M. Page and M. R. Jeffords, *Ill. Nat. Hist. Surv. Bull.* 34, art. 4, 357–477.

Burrill, T. J. 1881. Blight, or bacteria ferments in fruit trees. *Ind. Hort. Soc. Trans.* 20:84–91.

Burrill, T. J., and S. A. Forbes. 1885. Report on agricultural experiment stations. *Trans. Ill. Hort. Soc.* 18:52–71.

Bynum, W. F. 1994. *Science and the Practice of Medicine in the Nineteenth Century*. Cambridge: Cambridge University Press.

Caesar, J. 1980. *The Battle for Gaul*. Boston: David R. Godine.

Cairns, J., and J. R. Pratt. 1993. 2. A history of biological monitoring using benthic macroinvertebrates. In *Freshwater Biomonitoring and Benthic Macroinvertebrates*, edited by D. M. Rosenberg and V. H. Resh, 10–27. New York: Chapman and Hall.

Cameron, J. W. M. 1973. Insect pathology. In *History of Entomology*, edited by R. F. Smith, T. E. Mittler, and C. N. Smith, 285–306. *Ann. Rev.*, Palo Alto, Calif.

Carlander, K. D., R. S. Campbell, and W. H. Irwin. 1963. Mid-continent states. Chap. 11 in *Limnology in North America*, edited by D. G. Frey, 317–348. Madison: University of Wisconsin Press.

Carter, J. C. Applied botany and plant pathology. In Mills et al. 1959. "A Century of biological research." *Ill. Nat. Hist. Surv. Bull.* 27, art. 2, 145–62.

Cassedy, J. H. 1991. *Medicine in America: A Short History*. Baltimore: Johns Hopkins University Press.

Catton, B. 1956. *This Hallowed Ground: The Story of the Union Side of the Civil War*. Garden City, N.Y.: Doubleday.

Chapman, H. H. and R. B. Miller. 1924. Second report on a forest survey of Illinois: The economics of forestry in the state. *Ill. Nat. Hist. Surv. Bull.* 15, art. 3, 46–172.

Clayton, J. 1970. *The Illinois Fact Book and Historical Almanac, 1673–1968*. Carbondale and Edwardsville: Southern Illinois University Press.

Coleman, W. 1971. *Biology in the Nineteenth Century*. New York: Wiley and Sons.

Commager, H. S. 1950. *The American Mind: An Interpretation of American Thought and Character Since the 1880s*. New Haven: Yale University Press.

Connelly, T. L. 1971. *Autumn of Glory: The Army of Tennessee, 1862–1865*. Baton Rouge: Louisiana State University Press.

Cook, J. W., and J. V. McHugh. 1882. *A History of the Illinois State Normal University, Normal, Illinois*. Bloomington: Pantagraph Printing.

Darrah, W. C. 1969. *Powell of the Colorado*. Princeton, N.J.: Princeton University Press.

Davenport, E. A., ed. 1888. *History of the Ninth Regiment Illinois Cavalry Volunteers*. Chicago: Donahue and Henneberry.

Davenport, F. G. 1958. Natural scientists and the farmers of Illinois, 1865–1900. *J. Ill. St. Hist. Soc.* 51(4):357–79.

Davidson, R. H., and W. F. Lyon. 1987. *Insect Pests of Farm, Garden, and Orchard*. New York: John Wiley and Sons.

DeBary, A. 1887. *Vorlesungen über Bacterien*. 2d ed. Leipzig: Wilhelm Englemann.

DeBow, J. D. 1953. *The Seventh Census of the United States*. 1850. Washington, D.C.

Decker, G. A. 1958. Economic entomology. In Mills et al. 1959. "A Century of biological research." *Ill. Nat. Hist. Surv. Bull.* 27, art. 2, 104–26.

Dexter, R. W. 1973. Historical sketch of Agassiz's summer school on Penikese Island—The Anderson School of Natural History, 1873–1973. *In* Centennial Commemoration of the Founding of the Anderson School of Natural History, 2 pp.

Drude, O. 1906. The position of ecology in modern science. In *Congress of Arts and Sciences, Universal Exposition, St. Louis, 1904*. Vol. 5, *Biology, Anthropology, Psychology, Sociology*, edited by H. J. Rogers, 179–90. Boston: Houghton Mifflin.

Dunlap, T. R. 1999. *Nature and the English Diaspora: Environment and History in the United States, Canada, Australia, and New Zealand*. Cambridge: Cambridge University Press.

Dupree, A. H. 1957. *Science in the Federal Government: A History of Policies and Activities to 1940*. Cambridge: Belknap Press of Harvard University Press.

East, B. B. 1958. Former technical employees. In Mills et al. 1959. "A Century of biological research." *Ill. Nat. Hist. Surv. Bull.* 27, art. 2, 215–18.

Egerton, F. N. 1977. Ecological studies and observations before 1900. In *History of American Ecology*, edited by F. N. Egerton, 311–72. New York: Arno Press.

———. 1983. History of ecology: Achievements and opportunities. Pt. 1. *J. Hist. Biol.* 16(2):259–310.

Elson, H. W. 1911. *The Photographic History of the Civil War*. Vol. 2, *Two Years of Grim War*. New York: Review of Reviews.

Essig, E. O. 1931. *A History of Entomology*. New York: Macmillan.

Farber, P. L. 1982. Discussion paper: The transformation of natural history in the nineteenth century. *J. Hist. Biol.* 15(1):145–52.

Farquharson, A. S. L. 1944. *Marcus Aurelius: Meditations*. New York: Everyman's Library, Knopf.

Faust, P. L., ed. 1991. *Historical Times Illustrated Encyclopedia of the Civil War*. New York: Harper Collins.

Federal Writers Project. 1947. *Illinois: A Descriptive and Historical Guide*. Chicago: A. C. McClurg.

Fiske, J. 1901. *The Mississippi Valley in the Civil War*. Boston: Houghton Mifflin.

Foote, S. 1974. *The Civil War: A Narrative. Red River to Appomattox*. New York: Random House.

Forbes, E. B. 1930. Stephen Alfred Forbes: His Ancestry, Education, and Character. In *In Memoriam, Stephen Alfred Forbes, 1844–1930*, 5–15.

Forbes, R. M. 1989. Ernest Browning Forbes, 1876–1966: A brief biography. *J. Anim. Sci.* 67:849–52.

Forbes, S. A. 1870. New Plants—*Saxifraga forbesii* (n.sp.) and a specimen of *Heuchera. Am. Entomol. and Bot.* 2(9):288.

———. 1872. The independent study of natural history. *Chic. Schoolmaster* 5(54):313–16.

———. 1873. Natural history in the public schools. *Ill. Schoolmaster* 6(66):363–73.

———. 1874. Direction for collecting and preserving specimens of natural history. Prepared for the School and College Association of Natural History of the State of Illinois, Normal, Ill. 11 pp.

———. 1874–75. Suggestions to teachers of zoology. *Ill. Schoolmaster* 7:391–395, 8:73–79.

———. 1876. List of Illinois Crustacea, with descriptions of new species. *Bull. Ill. Mus. Nat. Hist.* 1(1):3–25.

———. 1878. The food of Illinois fishes. *Bull. Ill. St. Lab. Nat. Hist.* 1(2):71–89.

———. 1880. The food of birds. *Bull. Ill. St. Lab. Nat. Hist.* 1(3):80–148.

———. 1880. The food of birds: the thrush family. *Trans. Ill. St. Hort. Soc.* 13:120–72.

———. 1880. The food of fishes. *Bull. Ill. St. Lab. Nat. Hist.* 1(3):18–65.

———. 1880. The food of the bluebird (*Sialia sialis* L.). *Am. Entomol.* 1 (n.s.) (9):215–18, and (10):231–34.

———. 1880. On some interactions of organisms. *Bull. Ill. St. Lab. Nat. Hist.* 1(3):3–17.

————. 1880. On the food of young fishes. *Bull. Ill. St. Lab. Nat. Hist.* 1(3):66–79.

————. 1881. Supplementary report on the food of the thrush family. *Trans. Ill. St. Hort. Soc.* 14:106–26.

————. 1882. Bacterium: a parasite of the chinch bug. *Am. Nat.* 16:824–25.

————. 1882. On some entomostraca of Lake Michigan and adjacent waters. *Am. Nat.* 16(7):537–42, and (8)640–49.

————. 1882. The ornithological balance-wheel. *Trans. Ill. St. Hort. Soc.* 15:120–31.

————. 1882. The regulative action of birds upon insect oscillations. *Bull. Ill. St. Lab. Nat. Hist.* 1(6):3–32.

————. 1882. The State Laboratory of Natural History. In J. W. Cook and J. V. McHugh, *A History of the Illinois State Normal University, Normal, Illinois,* 236–49, Bloomington.

————. 1883. The first food of the common white-fish. *Bull. Ill. St. Lab. Nat. Hist.* 1(6):95–109.

————. 1883. The food of the smaller fresh-water fishes. *Bull. Ill. St. Lab. Nat. Hist.* 1(6):65–94.

————. 1883. Insects affecting corn. *Ill. St. Entomol. Off. Circ.* 1–21.

————. 1883. Notes on economic ornithology. *Trans Ill. St. Hort. Soc.* 16:58–71.

————. 1884. On a contagious disease of caterpillars. *Trans. Ill. St. Hort. Soc.* 17:29–41.

————. 1884. On the life-histories and immature stages of three Eumolpini. *Psyche* 4:123–30, 167–68.

————. 1885. A season's work in horticultural entomology. *Trans. Ill. St. Hort. Soc.* 18:117–27.

————. 1886. The chinch-bug in Illinois. *Ill. St. Entomol. Off. Circ.* 8 pp. (September).

————. 1886. Experiments on the codling moth and curculios. *Trans. Ill. St. Hort. Soc.* 19:103–24.

————. 1886. Studies on the contagious diseases of insects. *Bull. Ill. St. Lab. Nat. Hist.* 2:257–321.

————. 1887. Arsenical poisons for the codling moth—record and discussion of experiments for 1885 and 1886. *Trans. Ill. St. Hort. Soc.* 20:109–19.

————. 1887. On the chinch-bug in Illinois: present conditions and prospects for 1887 and 1888; remedial procedure recommended. *Ill. St. Entomol. Off. Bull.* 2:27–43.

————. 1888. Food of the fishes of the Mississippi Valley. Paper read to the American Fisheries Society 17th annual meeting, Detroit, May 16. 17 pp.

————. 1888. Note on chinch-bug diseases. *Psyche* 5:110–11.

————. 1888. On the food relations of freshwater fishes: a summary and discussion. *Bull. Ill. St. Lab. Nat. Hist.* 2(8):475–538.

————. 1888. On the present state of our knowledge concerning contagious insect diseases. *Psyche* 5:3–22.

————. 1890. The Hessian fly. *Ill. St. Bd. Agric. Crop Rep. Circ.* 3 pp. (June).

————. 1890. Preliminary report upon the invertebrate animals inhabiting Lakes Geneva and Mendota, Wisconsin, with an account of the fish epidemic in Lake Mendota in 1884. *U. S. Fish Comm. Bull.* 8:473–87.

————. 1890. Synopsis of recent work with arsenical insecticides. *Trans. Ill. St. Hort. Soc.* 23:310–325.

————. 1891. On a bacterial insect disease. *North Amer. Pract.* 3:401–5.

————. 1891. On some Lake Superior Entomostraca. *U. S. Fish Comm. Rept. for 1887–88*, pp. 701–18.

————. 1891. Report of progress in economic entomology. *U. S. Off. Agric. Exp. Stn. Bull.* 3:29–34.

————. 1892. Bacteria normal to digestive organs in Hemiptera. *Bull. Ill. St. Lab. Nat. Hist.* 4(1):1–7.

————. 1893. On a bacterial disease of the chinch-bug. Proc. of the 12th and 13th annual meetings of the Society for Promotion of Agricultural Science, 1891–92, pp. 44–48.

————. 1893. A preliminary report on the aquatic invertebrate fauna on the Yellowstone National Park, Wyoming, and of the Flathead region of Montana. *Bull. U. S. Fish Comm.* 11:207–58.

————. 1893. Presidential address; Association of Economic Entomologists, Aug. 14, 1893, Madison, Wis. *Insect Life* 6:61–70.

————. 1894. The aquarium of the United States Fish Commission at the World's Columbian Exposition, Report of the Director. *Bull. U.S. Fish Comm.* 13:143–58.

————. 1894. Damage to food-fish in Wabash River. U. S. 53d Congress, 2d Session, *H. R. Misc. Doc.* 196:2–4.

————. 1894. The spontaneous occurrence of white muscardine among chinch-bugs in 1895. 18th Report, St. Entomol. of Illinois, 75–78.

————. 1895. On contagious disease in the chinch-bug *Blissus leucopterus* Say). 19th Report, St. Entomol. of Illinois, 16–18.

————. 1896. Special Report of the Biological Experiment Station. In 18th Report, Board of Trustees, University of Illinois, 302–26.

————. 1897. On the principal corn insects and methods of controlling them. *Ill. Agriculturist*, pp. 36–48.

————. 1897. The San José scale in Illinois. *Univ. Ill. Agric. Exp. Stn. Bull.* 48:413–28.

————. 1898. Note on a new disease of the army worm. 20th Report, St. Entomol. of Illinois, 106–9.

————. 1898. The season's campaign against the San José and other scale insects in Illinois. *Trans. Ill. St. Hort. Soc.* 31:105–19.

————. 1899. Lessons from the year's work with the San José scale. *Trans. Ill. St. Hort. Soc.* 32:50–61.

————. 1900. The economic entomology of the sugar beet. 21st Report, St. Entomol. of Illinois, pp. 49–184.

————. 1900. The workings of the San José scale law. *Trans. Ill. St. Hist. Soc.* 33:150–59.

————. 1901. Annual statement concerning operations under the horticultural inspection act. *Trans. Ill. St. Hort. Soc.* 34:213–15.

————. 1901. Nursery inspection and orchard insecticide work in Illinois. *U.S. Off. Agric. Exp. Stn. Bull.* 99:173–76.

————. 1902. The corn bill-bugs in Illinois. *Univ. Ill. Agric. Exp. Stn. Bull.* 79:435–61.

————. 1902. Experiments with insecticides for the San José scale. *Univ. Ill. Agric. Exp. Stn. Bull.* 71:241–64.

————. 1903. Experiments with summer washes for the San José scale. 22d Report, St. Entomol. of Illinois, 96–97.

————. 1903. Report of inspection and insecticide operations, and the San José scale and other scale insects. *Trans. Ill. St. Hort. Soc.* 36:120–26.

————. 1904. The kind of economic entomology which the farmer ought to know. *Ill. St. Entomol. Off.* Pamphlet. 16 pp.

————. 1905. Field experiments and observations on insects injurious to Indian corn. *Univ. Ill. Agric. Exp. Stn. Bull.* 104. 8 pp. (abstract).

————. 1905. Insects injurious to corn. A Conference on the Corn Insects of Illinois. 10th Ann. Meeting, Illinois Farmer's Institute, February 21, 1905, pp. 3–23.

————. 1905. Practical treatment of the San José scale. *Univ. Ill. Agric. Exp. Stn. Circ.* 85. 4 pp.

————. 1905. Report of nursery inspections and insecticide operations. *Trans. Ill. St. Hort. Soc.* 38:108–14.

————. 1906. Comparative experiments with various insecticides for the San José scale. *Univ. Ill. Agric. Exp. Stn. Bull.* 107. 3 pp. (abstract).

————. 1906. Report of nursery inspection and insecticide operations. *Trans. Ill. St. Hort. Soc.* 39:132–35.

————. 1906. Spraying apples for the plum curculio. *Univ. Ill. Agric. Exp. Stn. Bull.* 108:265–86.

————. 1907. Grierson's cavalry raid. *Trans. Ill. St. Hist. Soc.* 8th Annual Mtg., January 24–25, 1907. Oral paper. Pt. 2, publ. no. 12 of the Illinois State Historical Library, Springfield, 99–130.

————. 1907. History of the former state natural history societies of Illinois. *Science* 26(678):892–98.

————. 1907. On the local distribution of certain Illinois fishes: an essay in statistical ecology. *Bull. Ill. St. Lab. Nat. Hist.* 7, art. 8, 273–303.

————. 1907. Report of nursery inspection and orchard insecticide operations. *Trans. Ill. St. Hort. Soc.* 40:124–28, 131–32.

————. 1908. Practical treatment for the San José scale. *Ill. St. Entomol. Off. Circ.*, 4 pp. (February).

————. 1908. Report of nursery inspection and orchard insecticide operations. *Trans. Ill. St. Hort. Soc.* 41:196–201.

————. 1909. Aspects of progress in economic entomology. *J. Econ. Entomol.* 2(1):25–35.

————. 1909. The general entomological ecology of the Indian corn plant. *Am. Nat.* 43:286–301.

————. 1909. Report on nursery inspection and orchard insecticide operations. *Trans. Ill. St. Hort. Soc.* 42:63–66.

————. 1909. Symposium: The Illinois State Laboratory of Natural History and the Illinois State Entomologist's Office. *Trans. Ill. St. Acad. Sci.* 2:34–67.

———. 1910. Biological investigations on the Illinois River. *Ill. St. Lab. Nat. Hist.* Pamphlet. 14 pp.

———. 1910. Recent work of the entomologist's office. *Ill. Farm Inst. Rep.* 15:140–50.

———. 1911. The chinch-bug abroad again in Illinois. *Ill. St. Entomol. Off. Circ.* 4 pp. (May).

———. 1911. Illinois Biological Station. *Int. Rev. Hydrobiol.* 4:226–27.

———. 1911. Report on nursery inspection and orchard insecticide operations. *Trans. Ill. St. Hort. Soc.* 44:90–93.

———. 1911. War as an education. *The Illinois* 3(1):1–10.

———. 1912. Chemical and biological investigations on the Illinois River, midsummer of 1911. *Ill. St. Lab. Nat. Hist.* Pamphlet. 9 pp.

———. 1912. The chinch-bug situation in Illinois: plans for a cooperative campaign. *Univ. Ill. Agric. Exp. Stn. Circ.* 7 pp.

———. 1912. The native animal resources of the state. *Trans. Ill. Acad. Sci.* 5:37–48.

———. 1912. The 1912 chinch-bug campaign in Illinois. *Ill. St. Entomol. Off. Circ.* 17 pp. (November).

———. 1912. On black flies and buffalo-gnats (*Simulium*) as possible carriers of pellagra in Illinois. 27th Report, St. Entomol. of Illinois, 21–55.

———. 1913. Biological and chemical conditions on the upper Illinois River. *Ill. Water Supply Assoc. Proc.* 5:161–70.

———. 1913. The corn root aphis in Illinois. *Univ. Ill. Agric. Exp. Stn. Circ.* 7 pp. (January).

———. 1913. The *Simulium*-pellagra problem in Illinois. *Science* 37(942):86–91.

———. 1914. The chinch-bug situation: present prospects and practical plans. *Ill. St. Entomol. Off. Circ.* 3 pp. (February).

———. 1914. Freshwater fishes and their ecology. *Ill. St. Lab. Nat. Hist.* Pamphlet. 19 pp.

———. 1915. The ecological foundations of applied entomology. *Ann. Entomol. Soc. Am.* 8(1):1–19.

———. 1915. The insect, the farmer, the teacher, the citizen, and the state. *Ill. St. Lab. Nat. Hist.* Pamphlet. 14 pp.

———. 1915. Observations and experiments on the San José scale. *Univ. Ill. Agric. Exp. Stn. Bull.* 180:545–61.

———. 1916. The chinch-bug outbreak of 1910 to 1915. *Univ. Ill. Agric. Exp. Stn. Circ.* 189:3–59.

———. 1916. Report on nursery and orchard inspection. *Trans. Ill. St. Hort. Soc.* 49:59–66.

———. 1916. Thomas Jonathan Burrill (obituary). *Ill. Univ. Quart.* 1:409–17.

———. 1919. Recent forest survey of Illinois. *Trans. Ill. St. Hort. Soc.* 52:103–10.

———. 1921. Streams pollution in the Illinois basin. *Ill. Sportsman* 6(10):1–2.

———. 1922. The humanizing of ecology. *Ecology* 3(2):89–92.

———. 1924. Sewage pollution of the Illinois River. *Outdoor America* 3(5):35–36.

———. 1925. The lake as a microcosm. *Ill. Nat. Hist. Surv. Bull.* 15, art. 9, 537–50. Reprinted from *Bull. Sci. Assoc.*, 1887, Peoria, Ill., 77–87.

————. 1926. Recent progress of the forestry movement. *Ill. Blue Book for 1925–26*, 468–72.

————. 1928. The biological survey of a river system—its objects, methods, and results. *Ill. Nat. Hist. Surv. Bull.* 17, art. 7, 277–84.

————. 1928. The effects of stream pollution on fishes and their food. *Ill. Nat. Hist. Surv.* Pamphlet. 13 pp.

————. 1929. Concerning certain ecological methods of the Illinois Natural History Survey. *Trans. Ill. St. Acad. Sci.* 21:19–25.

————. 1930. State Natural History Survey. *Ill. Blue Book for 1929–30*, 393–404.

————. 1930. What the Natural History Survey is doing for the high school. *Ill. Nat. Hist. Surv.* Pamphlet. 11 pp.

————. 1977. *Ecological investigations of Stephen Alfred Forbes*. New York: Arno Press.

Forbes, S. A., and A. O. Gross. 1921. The orchard birds of an Illinois summer. *Ill. Nat. Hist. Surv. Bull.* 14, art. 1, 1–8.

————. 1922. The numbers and local distribution in summer of Illinois land birds of open country. *Ill. Nat. Hist. Surv. Bull.* 14, art. 6, 187–218.

————. 1923. On the numbers and local distribution of Illinois land birds of the open country in winter, spring, and fall. *Ill. Nat. Hist. Surv. Bull.* 14, art. 10, 397–453.

Forbes, S. A. and R. B. Miller. 1920. Concerning a forestry survey and forester for Illinois. *Ill. Nat. Hist. Surv. For. Circ.* 1. 7 pp.

Forbes, S. A., and R. E. Richardson. 1908. *The Fishes of Illinois. Ill. Nat. Hist. Surv. Rept.* 3, cxxxi + 357 pp.

————. 1913. Studies on the biology of the upper Illinois River. *Bull. Ill. St. Lab. Nat. Hist.* 9, art. 10, 481–574.

————. 1919. Some recent changes in Illinois River biology. *Ill. Nat. Hist. Surv. Bull.* 13, art. 6, 139–56.

Frey, D. G., ed. 1963. *Limnology in North America*. Madison: University of Wisconsin Press.

Fulwider, A. L. 1910. *History of Stephenson County, Illinois. A Record of Its Settlement, Organization, and Three-quarters of a Century of Progress*. Vol. 1. Chicago: S. J. Clarke.

Garland, H. 1899. *Boy Life on the Prairie*. New York: Macmillan.

Garman, W. H. 1890. A preliminary report on the animals of the Mississippi bottoms near Quincy, Illinois in August, 1888. Pt 1. *Bull. Ill. St. Lab. Nat. Hist.* 3(9):123–84.

Garraty, J. A. 1968. *The New Commonwealth, 1877–1890*. New York: Harper and Row.

Goldschmidt, R. B. 1915. Biographical Memoir of Charles Atwood Kofoid, 1865–1947. *Biogr. Mem. Natl. Acad. Sci.* 26:121–51.

Grant, U. S. 1990. *Memoir and Selected Letters: Personal Memoirs of U. S. Grant, Selected Letters, 1839–1865*. New York: Library of America.

Gunning, G. E. 1963. Illinois. Chap. 5 in *Limnology in North America*, edited by D. G. Frey, 163–89. Madison: University of Wisconsin Press.

Hagan, W. T. 1958. *The Sac and Fox Indians*. Norman: University of Oklahoma Press.

Haslam, S. M. 1990. *River Pollution: An Ecological View*. London: Belhaven Press.

Havera, S. P., and K. E. Roat. 1989. Forbes Biological Station: the past and the promise. *Ill. Nat. Hist. Surv. Spec. Publ.* 10:1–24.

Hays, R. G. 1980. *State Science in Illinois: The Scientific Surveys, 1850–1978*. Carbondale: Southern Illinois University Press.

Hendrickson, W. B. 1963. The forerunners of the Illinois State Academy of Science. *Trans. Ill. St. Acad. Sci.* 56(3):105–28.

Henry, R. S. 1944. *First With the Most: Forrest*. Indianapolis: Bobbs-Merrill.

Hesseltine, W. B. 1930. *Civil War Prisons: A Study in War Psychology*. Columbus: Ohio State University Press.

Hood, J. B. 1959. *Advance and Retreat: Personal Experiences in the United States and Confederate States Armies*. 2d ed. Bloomington: Indiana University Press.

Horn, S. F. 1952. *The Army of Tennessee*. Norman: University of Oklahoma Press.

———. 1968. *The Decisive Battle of Nashville*. Knoxville: University of Tennessee Press.

Horrell, C. W., H. D. Piper, and J. W. Voigt. 1973. *Land between the Rivers: The Southern Illinois Country*. Carbondale: Southern Illinois University Press.

Howard, L. O. 1930. *A History of Applied Entomology*. *Smiths. Misc. Coll.* 84:1–564.

———. 1932. Biographical Memoir of Stephen Alfred Forbes, 1844–1930. *Biogr. Mem. Natl. Acad. Sci.* 15:1–54.

Hurst, J. 1993. *Nathan Bedford Forrest: A Biography*. New York: A. A. Knopf.

Hutchinson, G. E. 1963. The prospect before us. Chap. 26 in *Limnology in North America*, edited by D. G. Frey, 683–90. Madison: University of Wisconsin Press.

Hynes, H. B. N. 1971. *The Biology of Polluted Waters*. Toronto: University of Toronto Press.

Johnson, E. L. 1981. Misconceptions about the early land-grant colleges. *J. Higher Educ.* 52(4):334–351.

Johnston, W. J. 1854. Sketches of the history of Stephenson County, Illinois, and incidents connected with the early settlement of the northwest. Reprint, publ. 30 of the Illinois State Historical Library, *Trans. Ill. St. Hist. Soc.* 1923, 217–320.

Jordan, D. S. 1878. A catalog of the fishes of Illinois. *Bull. Ill. St. Lab. Nat. Hist.* 1(2):37–70.

———. 1922. *The Days of a Man: Being Memories of a Naturalist, Teacher, and Minor Prophet of Democracy*. 2d vol. Yonkers, N. Y.: World Book.

Juday, C. 1908. Some aquatic invertebrates that live under anaerobic conditions. *Trans. Wis. Acad. Sci. Arts. Lett.* 16:10–16.

———. 1921. Quantitative studies of the bottom fauna in the deeper waters of Lake Mendota. *Trans. Wis. Acad. Sci. Arts Lett.* 20:461–93.

Kaiser, L. M. 1975. Wild eagle flight: A Union soldier's view of the Civil War and the South, 1861–1865. *La. Studies* 14:395–412.

Kerfoot, W. C., ed. 1980. *Evolution and Ecology of Zooplankton Communities*. Spec. Sympos. Vol. 3, ASLO. Hanover, N. H.: University Press of New England.

Kingsland, S. E. 1985. *Modeling Nature: Episodes in the History of Population Ecology*. Chicago: University of Chicago Press.

Kinley, D. 1930. Address. In *In Memoriam, Stephen Alfred Forbes, 1844–1930*, 32–35.

Kofoid, C. A. 1897. On some important sources of error in the plankton method. *Science* 6(153):829–32.

————. 1898. Plankton studies. I. Methods and apparatus in use in plankton investigations at the Biological Experiment Station of the University of Illinois. *Bull. Ill. St. Lab. Nat. Hist.* 5(1):1–25.

————. 1898. The fresh-water biological stations of America. *Am. Nat.* 32:391–406.

————. 1905. Illinois River plankton. *Science* 21(528):233–34.

————. 1908. Plankton studies. V. The plankton of the Illinois River, 1894–1899. Pt. 2. Constituent organisms and their seasonal distribution. *Bull. Ill. St. Lab. Nat. Hist.* 8, art. 1, 3–360.

Kohlstedt, S. G. 1988. 1. Museums on campus: A tradition of inquiry and teaching. In *The American Development of Biology*, edited by R. Rainger, K. R. Benson, and J. Maienschein, 1–1147. Philadelphia: University of Pennsylvania Press.

Kolkwitz, R., and M. Marsson. 1909. Oekologie der tierischen Saprobien. *Int. Rev. Hydrobiol.* 2:126–52.

Lanier, R. S., ed. 1987. *The Photographic History of the Civil War.* Vol. 5, *The Armies and Leaders*. Secaucus, N.J.: Blue and Grey Press.

Leckie, W. H., and S. A. Leckie. 1984. *Unlikely Warriors: General Benjamin H. Grierson and His Family*. Norman: University of Oklahoma Press.

Lewontin, R. C. 1983. The corpse in the elevator. *New York Review of Books* 24 (January 20), 34–37.

Lohmann, H. 1901. Ueber das Fischen mit Netzen aus Müllergaze Nr. 20 zu dem Zwecke quantitativen Untersuchungen des Auftriebs. *Wiss. Meeresunters.* Abteilung Kiel, Neue Folge 5:46–66.

————. 1908. Untersuchungen zur Feststellung des volkständigen Gehalts des Meeres an Plankton. *Wiss. Meeresunters.* Abteilung Kiel. Neue Folge 10:129–270.

Long, E. B. 1971. *The Civil War Day by Day: An Almanac, 1861–1865.* Garden City, N.Y.: Doubleday.

Longacre, E. G. 1972. *From Union Stars to Top Hat: A Biography of the Extraordinary General James Harrison Wilson*. Harrisburg, Pa.

Ludmerer, K. M. 1985. *Learning to Heal: The Development of American Medical Education.* New York: Basic Books.

Lynne, D. M. 1955. Wilson's cavalry at Nashville. *Civil War Hist.* 1:141–59.

Mallis, A. 1971. *American Entomologists*. New Brunswick, N.J.: Rutgers University Press.

Malone, D., ed. 1964. *Dictionary of American Biography*, vol. 10, New York: Charles Scribners and Sons.

Marshall, H. E. 1956. *Grandest of Enterprises: Illinois Normal University, 1857–1957*. Normal: Illinois St. Normal University.

McAtee, W. L. 1933. Economic ornithology. In *Fifty Years Progress of American Ornithology, 1883–1933*, 111–29, Lancaster, Pa.: American Ornithologists' Union.

McDonough, J. E., and T. L. Connelly. 1983. *Five Tragic Hours: The Battle of Franklin.* Knoxville: University of Tennessee Press.

McIntosh, R. P. 1985. *The Background of Ecology: Concept and Theory*. Cambridge: Cambridge University Press.

———. 1977. Pluralism in ecology. *Ann. Rev. Ecol. Syst.* 18:321–341.

McManis, D. 1964. The initial evaluation and utilization of the Illinois prairies, 1815–1840. University of Chicago, Dept. of Geography, Research Paper 94.

McMurray, R. M. 1982. *John Bell Hood and the War for Southern Independence.* Lexington: University Press of Kentucky.

McPherson, J. M., ed. 1989. *Battle Chronicles of the Civil War, 1864.* New York: Macmillan.

———. 1997. *For Cause and Comrades: Why Men Fought in the Civil War.* New York: Oxford University Press.

Metcalf, C. L. 1930. Obituary. Stephen Alfred Forbes, May 29, 1844–March 13, 1930. *Entomol. News* 41:175–78.

Metcalf, C. L., and W. P. Flint. 1962. *Destructive and Useful Insects: Their Habits and Control.* Revised by R. L. Metcalf. 4th ed. New York: McGraw-Hill.

Miller, F. T. ed. 1911. *The Photographic History of the Civil War.* Vol. 1, *The Opening Battles.* Springfield, Mass.: Patriot Publishing.

Mills, E. L. 1989. *Biological Oceanography: An Early History, 1870–1960.* Ithaca: Cornell University Press.

Mills, H. B. 1958. 1858–1958. In Mills et al. 1959. "A Century of biological research." *Ill. Nat. Hist. Surv. Bull.* 27, art. 2, 85–103.

Mills, H. B. 1964. Stephen Alfred Forbes. *Syst. Zool.* 13:208–14.

Mills, H. B., G. C. Decker, H. H. Ross, J. C. Carter, G. W. Bennett, T. G. Scott, J. S. Ayars, R. R. Warrick, and B. B. East. 1959. A century of biological research. *Ill. Nat. Hist. Surv. Bull.* 27, art. 2, 85–234.

Mills, H. B., W. C. Starrett, and F. C. Bellrose. 1966. Man's effect on the fish and wildlife of the Illinois River. *Ill. Nat. Hist. Surv. Bull. Notes* 57:1–24.

Möbius, K. A. 1877. Die Austern und die Austernwirtschaft. Berlin: Verlag von Wiegandt, Hempel & Parey. (Also as The Oyster and Oysterculture, in *U. S. Fish Comm. Report* for 1883, 683–751.)

Moores, R. G. 1970. *Fields of Rich Toil: The Development of the University of Illinois College of Agriculture.* Urbana: University of Illinois Press.

Morrison, S. E. 1972. *The Oxford History of the American People:* Vol. 2, *1789 through Reconstruction.* New York: New American Library.

Mumford, L. 1971. *The Brown Decades: A Study of the Arts in America, 1865–1895.* New York: Dover.

Nevins, A. 1917. *Illinois.* New York.

Nichols, S., and L. Entine. 1976. *Prairie Primer.* Madison: University of Wisconsin—Extension.

Norwood, W. F. 1970. Medical education in the U.S. before 1900. In *The History of Medical Education,* edited by C. D. O'Malley, 463–89. Berkeley: University of California Press.

Nyhart, L. K. 1995. *Biology Takes Form: Animal Morphology and the German Universities, 1800–1900.* Chicago: University of Chicago Press.

Pease, T. C. 1930. Stephen Alfred Forbes, 1844–1930: A necrology. *J. Ill. St. Hist. Soc.* 23(3):543–48.

Pennak, R. W. 1963. Rocky Mountain states. Chap. 12 in *Limnology in North America*, edited by D. G. Frey, 349–69. Madison: University of Wisconsin Press.

————. 1989. *Fresh-water Invertebrates of the United States: Protozoa to Mollusca*. 3d ed. New York: John Wiley and Sons.

Petulla, J. M. 1988. *American Environmental History*. Columbus, Ohio: Merrill.

Pierce, F. C. P. 1892. *Forbes and Forbush Genealogy: The Descendants of Daniel Forbush*. Chicago.

Pierce, Sgt. L. B. 1865. *The History of the Second Iowa Cavalry*. Burlington: Hawk-Eye Steam Book and Job Printing.

Pollak, B. S. 1963. The Peoria Scientific Association. *Peoria Hist. Soc.* no. 7602-1, 9 pp.

Pooley, W. V. 1908. *The Settlement of Illinois from 1830 to 1850*. Bulletin of the University of Wisconsin 220, History Series 1(4).

Purdy, W. C. 1916. Investigations of the pollution and sanitary condition of the Potomac watershed. Potomac plankton and environmental factors. *U. S. Publ. Health Serv. Hyg. Lab. Bull.* 104:130–91.

Rainger, R., K. R. Benson, and J. Mainschein, eds. 1988. *The American Development of Biology*. Philadelphia: University of Pennsylvania Press.

Ramsey, P., ed. 1957. *Jonathan Edwards, Freedom of the Will*. Vol. 1, *The Works of Jonathan Edwards*. New Haven: Yale University Press.

Reece, J. N. 1901. *History of Seventh Cavalry*. Report of the Adjutant General, State of Illinois. Vol. 8 for 1861–66. Springfield: State of Illinois.

Richardson, R. E. 1921. Changes in the bottom and shore fauna of the middle Illinois River and its connecting lakes since 1913–1915 as a result of the increase southward of sewage pollution. *Ill. Nat. Hist. Surv. Bull.* 14, art. 4, 33–75.

————. 1921. The small bottom and shore fauna of the middle and lower Illinois River and its connecting lakes, Chillicothe to Grafton: its valuation; its sources of food supply; and its relation to the fishery. *Ill. Nat. Hist. Surv. Bull.* 13, art. 15, 363–522.

————. 1925. Changes in the small bottom fauna of Peoria Lake, 1920 to 1922. *Ill. Nat. Hist. Surv. Bull.* 15, art. 5, 327–388.

————. 1925. Illinois River bottom fauna in 1923. *Ill. Nat. Hist. Surv. Bull.* 15, art. 6, 391–422.

————. 1928. The bottom fauna of the middle Illinois River, 1913–1925: its distribution, variation, and index value in the study of stream pollution. *Ill. Nat. Hist. Surv. Bull.* 17, art. 12, 387–475.

Riegel, R. E. 1949. *Young America, 1830–1840*. Westport, Conn.: Greenwood.

Riley, C. V. 1883. General notes, Entomology. The use of contagious germs as insecticides. *Am. Nat.* 17:1169–70.

Ripley, V. S. 1936. The settlement of Aux Plaines–now Riverside. Chap 4 in *Riverside Then and Now: History of Riverside, Illinois*, edited by H. J. Bassman, 46–66. Riverside, Ill.

Robbins, R. M. 1942. *Our Landed Heritage: The Public Domain, 1776–1936*. Princeton, N.J.: Princeton University Press.

Robertson, J. I. Jr. 1988. *Soldiers Blue and Gray*. Columbia: University of South Carolina Press.

Rodgers, A. D. 1944. *John Merle Coulter, Missionary in Science.* Princeton, N.J.: Princeton University Press.

Roe, D. A. 1973. *A Plague of Corn: The Social History of Pellagra.* Ithaca: Cornell University Press.

Rolston, L. H., and C. E. McCoy. 1966. *Introduction to Applied Entomology.* New York: Ronald.

Rosenberg, D. M., and V. H. Resh, eds. 1993. *Freshwater Biomonitoring and Benthic Invertebrates.* New York: Chapman and Hall.

Ross, H. H. 1958. Faunistic survey. In Mills et al. 1959. "A Century of biological research." *Ill. Nat. Hist. Surv. Bull.* 27, art. 2, 127–44.

Rothstein, W. G. 1972. *American Physicians in the Nineteenth Century: From Sects to Science.* Baltimore: Johns Hopkins University Press.

———. 1987. *American Medical Schools and the Practice of Medicine: A History.* New York: Oxford University Press.

Schneider, D. W. 1996. Enclosing the floodplain: resource conflict on the Illinois River, 1880–1920. *Env. Hist.* 1(2):70–96.

———. 2000. Local knowledge, environmental politics, and the founding of ecology in the United States: Stephen Forbes and the Lake as a Microcosm (1887). *Isis* 91(4): 681–705.

Scott, E. F. 1936. Forbes family letters and journals of Stephen Alfred Forbes, 1847–1870. One hundred years after the Forbes family settled in Illinois. Typescript.

Scott, T. G. 1958. Wildlife research. In Mills et al. 1959. "A Century of biological research." *Ill. Nat. Hist. Surv. Bull.* 27, art. 2, 179–201.

Scott, W. F. 1893. *Story of a Cavalry Regiment.* New York.

Shannon, F. A. 1945. *The Farmer's Last Frontier: Agriculture, 1860–1897.* New York: Farrar and Rinehart.

Shelford, V. E. 1938. The organization of the Ecological Society of America, 1914–19. *Ecology* 19(1):164–66.

Simon, J. Y. 1979. *The Papers of Ulysses S. Grant.* Vol. 8, *April–July 6, 1863.* Carbondale: Southern Illinois University Press.

Smith, E. F. 1911. *Bacteria in Relation to Plant Diseases.* Vol. 2. *Carnegie Inst. Publ.* 27. Washington, D.C.

Smith, F. 1926. Stephen Alfred Forbes: An appreciation. *Audubon Bull.* 17:19–25.

Smith, P. W. 1979. *The Fishes of Illinois.* Urbana: University of Illinois Press.

Smith, R. F., T. E. Mittler, and C. N. Smith, eds. 1973. *History of Entomology.* Palo Alto: Ann. Reviews.

Snow, F. H. 1895. Contagious diseases of the chinch-bug. 4th Ann. Rept. Dir. Univ. Kansas for 1894. 46 pp.

Solberg, W. V. 1968. *The University of Illinois, 1867–1894.* Urbana: University of Illinois Press.

Sparks, R. E., and W. C. Starrett. 1975. An electrofishing survey of the Illinois River, 1959–1974. *Ill. Nat. Hist. Surv. Bull.* 31, art. 8, 317–380.

Spencer, H. 1896. *The Principles of Biology.* Vol. 2. New York: Appleton.

Starr, S. Z. 1979. *The Union Cavalry in the Civil War.* Vol. 1, *From Fort Sumter to Gettysburg, 1861–1863.* Baton Rouge: Louisiana State University Press.

————. 1981. *The Union Cavalry in the Civil War.* Vol. 2, *The War in the East from Gettysburg to Appomattox, 1863–1865.* Baton Rouge: Louisiana State University Press.

————. 1985. *The Union Cavalry in the Civil War.* Vol. 3, *The War in the West, 1861–1865.* Baton Rouge: Louisiana State University Press.

Starrett, W. C. 1972. Man and the Illinois River. In *River Ecology and the Impact of Man,* edited by R. T. Oglesby, C. A. Carlson, and J. A. McCann, 131–69. New York: Academic Press.

Stauffer, R. C. 1957. Haeckel, Darwin, and ecology. *Q. Rev. Biol.* 32:138–44.

Steinhaus, E. A. 1949. *Principles of Insect Pathology.* New York: McGraw-Hill.

————. 1956. Microbial control—the emergence of an idea. A brief history of insect pathology through the nineteenth century. *Hilgardia* 26(2):107–57.

Stephenson County Farm Bureau. 1983. *Stephenson County, Illinois Plat Book.* Freeport: Hutchinson and Nesbit.

Surby, R. W. 1865. *Grierson's Raids and Hatch's Sixty-four Days' March, with Biographical Sketches, and the Life and Adventures of Chickasaw, the Scout.* Chicago: Rounds and James.

Sword, W. 1994. *Embrace an Angry Wind: The Confederacy's Last Hurrah, Spring Hill, Franklin, and Nashville.* Columbus, Ohio: General's Books.

Symonds, C. L. 1983. *A Battlefield Atlas of the Civil War.* Baltimore: Nautical and Aviation Publishing Company of America.

Taylor, B. 1883. *Hannah Thurston: Story of an American Life.* 2d ed. New York: G. P. Putnam's Sons.

Thomas, C. 1861. Plans for a natural history survey. *Trans. Ill. St. Agric. Soc.* 4:663–65

Thompson, H., ed. 1987. *The Photographic History of the Civil War.* Vol. 4, *Prisons and Hospitals.* Secaucus, N.J.: Blue and Grey Press.

Thompson and Everts. 1871. Combination atlas map of Stephenson County, Illinois. Compiled, drawn, and published from personal examinations and surveys. Freeport Public Library, Freeport, Ill.

Tilden, M. H. 1880. *The History of Stephenson County, Illinois. Containing a History of the County, Its Cities, Towns, etc.* Chicago: Western Historical.

Trelease, W. 1932. Biographical Memoir of John Merle Coulter, 1851–1928. *Biogr. Mem. Natl. Acad. Sci.* 14:97–123.

True, A. C. 1937. A history of agricultural experimentation and research in the United States, 1607–1925, including a history of the United States Department of Agriculture. Miscellaneous Publication 251, U.S. Dept. of Agriculture. Washington, D.C.: Government Printing Office.

Vago, C. 1963. Predispositions and Interrelations in Insect Diseases. In *Insect Pathology: An Advanced Treatise,* edited by E. A. Steinhaus, 1:339–379. New York: Academic Press.

Van Cleave, H. J. 1930. Stephen Alfred Forbes as a Scientist. In *In Memoriam, Stephen Alfred Forbes, 1844–1930,* 24–28.

Vasey, G. 1870. Plants to name—Mr. S. A. Forbes, Benton, Ill. *Am. Entomol. & Bot.* 2(8):256.

Ward, G. C. 1990. *The Civil War: An Illustrated History.* New York: Alfred C. Knopf.

Ward, H. B. 1930. Stephen Alfred Forbes—A Tribute. *Science* 71(1841): 378–81.

Warrick, R. R. "Library." In Mills et al. 1959. "A Century of biological research." *Ill. Nat. Hist. Surv. Bull.* 27, art. 2, 210–14.

Welcher, E. J. 1993. *The Union Army, 1861–1865: Organization and Operations.* Vol. 2, *The Western Theater.* Bloomington and Indianapolis: Indiana University Press.

Wiebe, R. H. 1967. *The Search for Order, 1877–1920.* New York: Hill and Wang.

Wilder, C. T. 1896. University of Illinois Biological Station. *The Illini* 25(20):314–20 and (21):331–39.

Wilhm, J. L. 1975. Biological indicators of pollution. Chap. 15 in *River Ecology,* edited by B. A. Whitton, 375–402. Berkeley: University of California Press.

Wilson, J. H. 1912. *Under the Old Flag: The Recollections of Military Operations in the War for the Union, The Spanish War, The Boxer Rebellion, etc.* Vol. 2. New York: D. Appleton.

Winsor, M. P. 1972. Stephen Alfred Forbes. *Dict. Sci. Biogr.* 5:69–71.

Wright, A. H., and A. A. Wright. 1950. Agassiz's address at the opening of Agassiz's Academy. *Am. Midl. Nat.* 43(2):503–6.

Index